Studies in Computational Intelligence

Volume 797

Series editor

Janusz Kacprzyk, Polish Academy of Sciences, Warsaw, Poland
e-mail: kacprzyk@ibspan.waw.pl

The series "Studies in Computational Intelligence" (SCI) publishes new developments and advances in the various areas of computational intelligence—quickly and with a high quality. The intent is to cover the theory, applications, and design methods of computational intelligence, as embedded in the fields of engineering, computer science, physics and life sciences, as well as the methodologies behind them. The series contains monographs, lecture notes and edited volumes in computational intelligence spanning the areas of neural networks, connectionist systems, genetic algorithms, evolutionary computation, artificial intelligence, cellular automata, self-organizing systems, soft computing, fuzzy systems, and hybrid intelligent systems. Of particular value to both the contributors and the readership are the short publication timeframe and the world-wide distribution, which enable both wide and rapid dissemination of research output.

More information about this series at http://www.springer.com/series/7092

Andrzej Obuchowicz

Stable Mutations for Evolutionary Algorithms

Springer

Andrzej Obuchowicz
Faculty of Computer Science, Electrical
and Control Engineering
Institute of Control and Computation
Engineering
University of Zielona Góra
Zielona Góra, Poland

ISSN 1860-949X ISSN 1860-9503 (electronic)
Studies in Computational Intelligence
ISBN 978-3-030-13184-5 ISBN 978-3-030-01548-0 (eBook)
https://doi.org/10.1007/978-3-030-01548-0

Softcover re-print of the Hardcover 1st edition 2019

This Springer imprint is published by the registered company Springer Nature Switzerland AG
The registered company address is: Gewerbestrasse 11, 6330 Cham, Switzerland

To my wife
Beata
and our children
Maria, Adam, and Tomasz

Contents

Abbreviations

DSM	Discrete spectral measure
EP	Evolutionary programming
$EP_{N;\alpha}$	EP with mutation based on $NS\alpha S(\sigma)$ distribution
$EP_{I;\alpha}$	EP with mutation based on $IS\alpha S(\sigma)$ distribution
$EP_{S;\alpha}$	EP with mutation based on $S\alpha SU(\sigma)$ distribution
ES	Evolutionary strategy
ESSS	Evolutionary search with soft selection
$ESSS_{N;\alpha}$	ESSS with mutation based on $NS\alpha S(\sigma)$
$ESSS_{I;\alpha}$	ESSS with mutation based on $IS\alpha S(\sigma)$
$ESSS_{S;\alpha}$	ESSS with mutation based on $S\alpha SU(\sigma)$
$ESSS_{\alpha}$-DM	ESSS with directional distribution $M(\mu, \kappa)$
ESTS	Evolutionary search with tournament selection
$ESTS_{N;\alpha}$	ESTS with mutation based on $NS\alpha S(\sigma)$
$ESTS_{I;\alpha}$	ESTS with mutation based on $IS\alpha S(\sigma)$
$ESTS_{S;\alpha}$	ESTS with mutation based on $S\alpha SU(\sigma)$
$ESTS_{\alpha}$-DM	ESTS with directional distribution $M(\mu, \kappa)$
GA	Genetic algorithm
GP	Genetic programming
i.i.d.	Independent and identical distribution
$(\eta+\lambda)$ ES	ES with a succession of the best η individuals from joined base and offspring populations
$(\eta+\lambda)$ $ES_{N;\alpha}$	$(\eta+\lambda)$ ES with mutation based on $NS\alpha S(\sigma)$
$(\eta+\lambda)$ $ES_{I;\alpha}$	$(\eta+\lambda)$ ES with mutation based on $IS\alpha S(\sigma)$
$(\eta+\lambda)$ $ES_{S;\alpha}$	$(\eta+\lambda)$ ES with mutation based on $S\alpha SU(\sigma)$
(η, λ) ES	ES with a succession of the best η individuals from λ offspring
(η, λ) $ES_{N;\alpha}$	(η, λ) ES with mutation based on $NS\alpha S(\sigma)$
(η, λ) $ES_{I;\alpha}$	(η, λ) ES with mutation based on $IS\alpha S(\sigma)$
(η, λ) $ES_{S;\alpha}$	(η, λ) ES with mutation based on $S\alpha SU(\sigma)$
SGA	Simple GA

Symbols

$\mathbb{R}$	Set of real numbers
$\mathcal{G}$	Genetic universe
$\mathcal{D}$	Phenotype universe
η	Number of individuals in a population
$P(t)$	Population in the t-th iteration
$\mathcal{P}_\eta(\mathcal{G})$	Set of all possible populations, i.e., multisets of η elements from $\mathcal{G}$
a_k^t	k-th individual from the population $P(t)$
$\Phi(\cdot)$	Fitness function
$A, B, \ldots, X, Y, Z$	Random variables
$A \overset{d}{=} B$	Random variables A and B are of the same distribution
$X_{i:\lambda}$	i-th random variable of the ordered statistics of λ variables
α	Stability index (tail index)
β	Skewness parameter
σ	Scale parameter
μ	Localization parameter
$\varphi(\cdot)$	Characteristic function
$p(\cdot)$	Probability density function
$E(\cdot)$	Expectation value
$Var(\cdot)$	Variance
$U(a,b)$	Continuous uniform distribution on (a, b)
$N(\mu, \sigma^2)$	Normal (Gaussian) distribution
$C(\mu, \sigma)$	Cauchy distribution
$Levy(\mu, \sigma)$	Lévy distribution
$S_\alpha(\sigma, \beta, \mu)$	α-stable distribution
$S\alpha S(\sigma)$	Symmetric α-stable distribution
Γ_s	Spectral measure
$\mathbf{N}(\mu, \mathbb{C})$	Multidimensional elliptic Gaussian distribution
$NS\alpha S(\sigma)$	Multidimensional non-isotropic symmetric α-stable distribution
$IS\alpha S(\sigma)$	Multidimensional isotropic α-stable distribution

$S\alpha SU(\sigma)$	Multidimensional isotropic distribution with an α-stable generator
$\mathscr{F}$	General search space
$MF(\mu, \kappa)$	Von Mises–Fisher distribution
$M(\mu, \kappa)$	Directional distribution

Chapter 1
Introduction

Practically any engineering activity, be it design, construction, modeling, control, etc., sooner or later leads to the necessity of solving a set of optimization problems. In practice, these problems usually appear to be multi-modal, sometimes multi-criteria or non-stationary (changing during the searching process). Therefore, standard optimization methods applied to solve them is inefficient. These techniques are usually based on the so-called *hard selection*—new base points for further searching space exploration are selected from the best points previously obtained. Using such procedures, the solution sequence is usually trapped near to the first found local extremum, without any possibility to localize others. So, the possibility of finding a global optimum is strongly limited in this case. Methods, known from the literature, which try to overcome this limitation can be divided into two classes: enumerative and stochastic ones.

The first group is mainly dedicated to discrete and finite spaces, and these methods usually consist in the examination of all solutions in order to designate the best one. These approaches are, of course, of low efficacy, especially in the case of high dimensional search spaces, where they cannot be realized in reasonable time, since a full review is impossible to be performed in reasonable time here. Thus, methods of heuristic search can be used (Schalkoff 1990). In the specific case, their effectiveness strongly depends on the chosen heuristic strategy of domain searching.

Stochastic optimization techniques are usually applied to multi-modal and multi-dimensional problems, for which the derivative is difficult or impossible to compute. The advantage of these algorithms is their simple computer implementation. However, they are time-consuming. In most of them, we can distinguish two phases: global search and local search. The former determines starting points for local search. Usually, these are randomly chosen from the domain. The goal of local search is to find the local extremum, which is the attractor for the given starting point. Among stochastic methods one can list several classes, e.g., pure random search (Monte-Carlo methods), multiple random starts, clustering methods, random direction methods, or search concentration methods (Birge and Louveaux 1997; Zieliński and Neumann 1983). The main disadvantage of stochastic methods lies in their

A. Obuchowicz, *Stable Mutations for Evolutionary Algorithms*,
Studies in Computational Intelligence 797,
https://doi.org/10.1007/978-3-030-01548-0_1

chaotic manner, which does not take into consideration information contained in previously evaluated points.

Particularly interesting stochastic methods of global extremum searching are meta-heuristic ones (Trojanowski 2008), like simulated annealing, evolutionary algorithms, particle swarm optimization and other algorithms based on the swarm idea, etc. Special attention in this book is paid to evolutionary algorithms; however, the presented results can be applied to other global optimization techniques in $\mathbb{R}^n$. The evolution is the natural way of development. Species acquire their properties and abilities by natural selection, seemingly a blind process, which allows mainly well-fitted individuals to survive and procreate. This mechanism permits transferring profitable features to next generations, thus we have some kind of 'intelligent' selection. But nature does not restrict itself to selecting only the best individuals in the population. Weakly fitted individuals have a chance to introduce their descendant to the next generation, too. The descendants are often gifted with attributes unknown in the current population, which can be useful in the future. Therefore, it is luring to introduce into optimization techniques the *soft selection* rule instead of the *hard* one, i.e., there is a possibility of choosing worse points as the base ones for further search. It appears that soft selection accelerates the probability of escaping from a local optimum trap.

Soft selection is one of the basic properties of evolutionary algorithms, which, thanks to this selection, are an effective computational intelligence technique dedicated to global optimization problems. The existing rich bibliography (Angeline and Kinnear 1996; Arabas 2001; Bäck 1995; Bäck et al. 1997; Davis 1987; Fogel 1998; Dasgupta and Michalewicz 1997; Galar 1990; Goldberg 1989; Holland 1992; Michalewicz 1996; Mitchel 1996; Schwefel 1995) confirms the thesis that a searching process based on natural evolution is a very robust and effective global optimization algorithm thanks to its adaptation abilities.

Most applications of evolutionary algorithms which use floating point representation of population individuals employ Gaussian mutation as a mutation operator (Bäck and Schwefel 1993; Fogel et al. 1966; Fogel 1995; Galar 1985; Michalewicz 1996; Rechenberg 1965). A new individual $\boldsymbol{x}$ is obtained by adding a normally distributed random value to each entry of a selected parent $\boldsymbol{y}$:

$$x_i = y_i + N(0, \sigma_i), \quad i = 1, \ldots, n. \tag{1.1}$$

The choice is usually justified by the central limit theorem. Mutations in nature are caused by a variety of physical and chemical influences that are not identifiable or measurable. These influences are considered to be independent and identically distributed (i.i.d.). The general central limit theorem states that the only non-trivial limit distribution of the random variable which is a normed sum of the series of i.i.d. random factors is some α-stable distribution (called also the Lévy-stable distribution or the heavy-tail distribution) (Fang et al. 1990; Nolan 2007; Samorodnitsky and Taqqu 1994). If the Lindeberg condition is obeyed, i.e., the first two absolute moments are finite, then the α-stable distribution reduces to the Gaussian one.

The main reason why α-stable distributions are not popular in the research community is the lack, in the general case, of the analytical form of the density function.

Only three distributions from the α-stable family possess the analytical form of the density function: the Gaussian, Cauchy and Lévy distributions. Fortunately, there are algorithmic formulas for simulation of random α-stable variables (Nolan 2007) as well as numerical algorithms for probability density functions, the cumulative distribution and quantile determination (Nolan et al. 2001).

The Cauchy distribution was the first one, apart from Gaussian, to be applied to the mutation operator in evolutionary algorithms (Bäck et al. 1997; Kappler 1996; Obuchowicz 2003a, b; Rudolph 1997; Yao and Liu 1996, 1997, 1999). However, there is an unequivocal definition of the one-dimensional Cauchy distribution and two of its multidimensional versions: isotropic and non-isotropic. The latter is constituted by a random vector composed of independent one-dimensional Cauchy variables (Fang et al. 1990; Obuchowicz 2003a). Thus, we can define non-isotropic Cauchy mutation by the substitution in (1.1) of the random variable the Gaussian distribution with the random variable of the Cauchy one. The course of the probability density function of the one-dimensional Cauchy distribution is similar to the Gaussian one, but the convergence of the Cauchy curve to the argument axis with the increasing of the absolute value of the argument to the infinity is much slower than in the case of the Gaussian one (α-stable distributions for low values of α are also called *heavy-tail distributions*). Thus, the probability of macro-mutations (mutations which locate the descendant far away from the basic point) significantly increases. Rudolph (1997) presented a theoretical analysis of the local convergence of simple evolutionary strategies $(1 + 1)$ES and $(1 + \lambda)$ES with Gaussian, isotropic and non-isotropic Cauchy mutations. He showed that the local convergence of evolutionary strategies with Gaussian and isotropic Cauchy mutations is very similar. Non-isotropic Cauchy mutation is characterized by significantly lower convergence to the local extremum. Yao and Liu (1996, 1997, 1999) show that evolutionary programming and evolutionary strategy algorithms with Cauchy mutation are very effective in comparison to those with standard Gaussian mutation in the case of a wide class of 30-dimensional global optimization tasks. This solution, however, has been obtained for benchmark objective functions of special form. Most of them (e.g., Griewank, Rastringin or Ackley functions—see Appendix B) possess local extrema localized in the nodes of some hyper-cubic net whose axes are parallel to the axis of the reference frame. These directions are significantly preferred by the multidimensional Cauchy distribution. The influence of reference frame selection on the effectiveness of evolutionary algorithms with Cauchy and Gaussian mutations in finding global optimization as well as the algorithms' ability of saddle crossing have been considered by Obuchowicz (2003a, b).

The suggestion that the application of the α-stable distribution instead of the Gaussian or Cauchy ones to evolutionary algorithms can be very attractive was first proposed by Gutowski (2001). However, he considered only some properties of α-stable distributions without any application to whatever evolutionary algorithm. Such application has been simultaneously and independently presented and analyzed by two research groups (Lee and Yao 2004; Obuchowicz and Prętki 2004a). Lee and Yao (2004) apply the evolutionary programming algorithm with non-isotropic α-stable mutations to solving a set of 14 global optimization problems. Obuchowicz and Prętki (2004a) study the influence of two little-known features—the so-called *symmetry effect* and *dead-surrounding effect*—of non-isotropic α-stable mutation on exploration and exploitation abilities of evolutionary algorithms. Hansen et al. (2006) notice some limitations of heavy-tail distributions by analyzing certain properties of isotropic and non-isotropic Cauchy distributions. They experimentally show that α-stable mutations are only effective in the optimization of the multi-modal function in research spaces of relatively low dimensions.

The goal of this book is to present a set of theoretical and experimental results which describe features of the wide family of α-stable distributions, various methods of their application to the mutation operator of evolutionary algorithms based on a real-number representation of the individuals, and, most of all, to equip these algorithms with features that enrich their effectiveness in solving multi-modal, multi-dimensional global optimization problems.

This book is divided into seven main chapters preceded by this introduction and supplemented at the end by a summary.

Basic terms of evolutionary algorithms, as well as their general outline, are introduced in Chap. 2. Next, classical versions of the best-known representative of evolutionary algorithms are described. Additionally, an evolutionary search with soft selection algorithm is presented. This simple evolutionary algorithm is one of the main algorithms used in simulating experiments described in further chapters of this monograph.

Definitions of the random stable variable, as well as the random stable vector, are included in Chap. 3. Moreover, foundations of the generator of pseudo-random variables of the α-stable distribution, as well as selected features of α-stable distributions, are presented. Special attention is focused on the features which can suggest that the application of α-stable distributions to the mutation operator of an evolutionary algorithm can increase the efficacy of the searching process. One of the most important theorems describes conditions of the existence of the expectation value of the first value of the ordered statistics of the α-stable random variables series (Obuchowicz and Prętki 2005). The main conclusion of this fact is that mutation based on the heavy-tail distribution can be more effective in the localization of the local extremum if only the parental individual has a possibility to create enough descendants. Chapter 3 contains also definitions of isotropic and non-isotropic random stable vectors, as well as a description of the isotropic distribution with the α-stable generator. The basic characteristics of the above-mentioned random vectors are discussed.

Non-isotropic stable mutations are the subject of Chap. 4. The analysis of the influence of the symmetry and dead surrounding effects on exploration and exploitation abilities of evolutionary search is emphasized.

Application of isotropic stable mutations, described in Chap. 5, allows avoiding problems with symmetry in a natural way. However, the dead surrounding effect still influences the searching process. It is surprising that the range of the dead surrounding decreases with the stability index α decreasing if isotropic stable mutation is used, unlike in the non-isotropic case. Further sections of that chapter contain an analysis of exploration and exploitation abilities of an evolutionary strategy with isotropic stable mutation as well as the robustness of the evolutionary algorithm to the inaccuracy of selection of mutation control parameters. An attempt at scale parameter adaptation strategy development for the chosen stability index is described in the last section of the chapter.

Implementations of multidimensional stable distributions in global optimization algorithms have so far been limited to two relatively simple techniques: the base is mutated by adding to it a random vector composed of independent random stable variables or an isotropic random stable vector. This limitation does not exploit many characteristics of α-stable distributions, which can turn out profitable during the global optimum searching process. In order to achieve the possibility of modeling more complicated dependencies between decision variables, the application of the discrete spectral measure to the generation of the wide class of random stable vectors is proposed (Obuchowicz and Prętki 2010). The main features of random stable vectors based on the discrete spectral measure are presented in Chap. 6. Using a simple simulating experiment it is shown that, for a given objective function, including dependencies between particular components of the random mutation vector increases the effectiveness of the extremum searching process. An attempt at the discrete spectral measure to the neighborhood of the current base point is shown at the end of the chapter.

In order to avoid inconveniences resulting from the symmetry and dead surrounding effects, a new type of mutation operator is proposed in Chap. 7. A descendant is obtained by a random vector with the α-stable generator added to the parent element. This random vector can be statistically decomposed into a directional random vector of a uniform distribution on the unit spherical surface and a α-stable random variable which represents the distance between the parent element and the descendant (Obuchowicz 2003b; Obuchowicz and Prętki 2004b). The obtained multidimensional distribution is isotropic, but, in the general case, it is not stable. The main advantage of the mutation considered is the minimization of the dead surrounding range, which is influenced only by the form of the objective function and is independent of the chosen mutation distribution. The analysis of the effectiveness of evolutionary algorithms with the mutation considered in local and global optimization tasks is given in Chap. 7. Results of the analysis of these algorithms' exploration and exploitation abilities are also presented.

Undoubtedly, isotropic probabilistic models possess both advantages and disadvantages. A low number of parameters needed for their definition is a very important advantage. Using them with stochastic optimization algorithms simplifies the searching process. A potential heuristic adaptation strategy of parameters regarding the current environment can be relatively simple. A low number of parameters is also somewhat disadvantageous, which can have negative influence on optimization efficacy. The isotropic probabilistic model allows taking into account the dependence between an objective function and the Euclidian distance from a base point only. If there exist stronger correlations between decision variables of the optimization problem, then a stochastic searching process with mutation based on the isotropic distribution can be extremely time consuming. In order to overcome these inconveniences, evolutionary algorithms with mutation based on the elliptic normal distribution have been proposed (Beyer and Schwefel 2002; Beyer and Arnold 2003; Hansen and Ostermeyer 2001). These algorithms are usually enriched with the mechanism of mutation parameter adaptation. Then it is possible to fit the exploration model to the currently explored optimization landscape. However, the mutation operator based on the elliptical normal distribution allows modeling the relationship between particular decision variables, but it possesses some important disadvantages. Both directions (the one leading to the expected improvement and the opposite one) have the same probability of selection. This fact leads to slow convergence of the searching processes considered. Another serious problem, called the *dimensionality curse*, is reflected in strong decreasing of the searching process with the increasing dimensionality of the searching space (Hansen et al. 2006; Prętki and Obuchowicz 2006). The choice of the spectral measure that prefers mutation in some improvement direction, which can be identified using information contained in distributions of previously verified populations in an evolutionary process, is an intuitive solution (Obuchowicz 2003b). Such directional distributions are proposed by Prętki and Obuchowicz (2006) and described in detail in Chap. 8.

Solutions set forth in this book for mutation operators based on α-stable distributions have already been successfully applied. Isotropic mutation with the α-stable generator, proposed in Chap. 7, has been successfully implemented in immune-based algorithms (Trojanowski and Wierzchon 2009), multi-swarm quantum particles (Trojanowski 2009) and differential evolution (Trojanowski et al. 2013) dedicated for dynamic optimization problems. Isotropic stable mutation, described in Chap. 5, has been applied the hierarchic genetic search system (HGS) (Obuchowicz 2015), especially in the hierarchic evolutionary inverse solver (Obuchowicz and Smołka 2016).

I would like to express my sincere thanks to a number of people. First of all, I am grateful to Professor Józef Korbicz for his continuous support and advice. I also wish to thank Professor Marek Gutowski for the inspiration for this research area, as well as Professors Jarosław Arabas, Roman Galar, Robert Schaefer and my son Adam for their active interest in my research and many stimulating suggestions. A significant part of results presented in this monograph have been obtained during several years of cooperation with Doctor Przemysław Prętki, who is a co-author of Chap. 5 of this book and to whom I hereby express my special thanks and hope that the experience gained during our cooperation will be useful in his professional career.

I also wish to express my deepest gratitude to my wife Beata for her everlasting patience, understanding and support during my work on this book and throughout all these years.

The research presented in this book has been partially funded by the National Science Centre in Poland through the project no. 2015/17/B/ST7/03704.

Zielona Góra, May 2018 *Andrzej Obuchowicz*

References

Angeline, P., & Kinnear, K. E. (1996). *Advances in genetic programming*. Cambridge: MIT Press.

Arabas, J. (2001). *Lectures on evolutionary algorithms*. Warsaw (in Polish): WNT.

Bäck, T. (1995). *Evolutionary algorithms in theory and practice*. Oxford: Oxford University Press.

Bäck, T., & Schwefel, H.-P. (1993). An overview of evolutionary algorithms for parameter optimization. *Evolutionary Computation, 1*(1), 1–23.

Bäck, T., Fogel, D. B., & Michalewicz, Z. (Eds.). (1997). *Handbook of evolutionary computation*. New York: Institute of Physics Publishing and Oxford University Press.

Beyer, H. G., & Schwefel, H. P. (2002). Evolution strategies-a comprehensive introduction. *Natural Computing, 1*(1), 3–52.

Beyer, H. G., & Arnold, D. V. (2003). Qualms regarding the optimality of cumulative path length control in CSA/CMA-evolution strategies. *Evolutionary Computation, 11*(1), 19–28.

Birge, J., & Louveaux, F. (1997). *Introduction to stochastic programming*. New York: Springer.

Dasgupta, D., & Michalewicz, Z. (Eds.). (1997). *Evolutionary algorithms for engineering applications*. Berlin: Springer.

Davis, L. (Ed.). (1987). *Genetic algorithms and simulated annealing*. San Francisco: Morgan Kaufmann.

Fang, K.-T., Kotz, S., & Ng, K. W. (1990). *Symmetric multivariate and related distributions*. London: Chapman and Hall.

Fogel, D. B. (1995). *Evolutionary computation: Toward a new philosophy of machine intelligence*. New York: IEEE Press.

Fogel, D. B. (1998). *Evolutionary computation: The fossil record*. New York: IEEE Press.

Fogel, L. J., Owens, A. J., & Walsh, M. J. (1966). *Artificial intelligence through simulated evolution*. New York: Wiley.

Galar, R. (1985). Handicapped individual in evolutionary processes. *Biological Cybernetics, 51*, 1–9.

Galar, R. (1990). *Soft selection in random global adaptation in R^n. A biocybernetic model of development*. Wrocław (in Polish): Technical University of Wrocław Press.

Goldberg, D. E. (1989). *Genetic algorithms in search, optimization and machine learning*. Addison-Wesley, Reading.

Gutowski, M. (2001). Lévy flights as an underlying mechanism for global optimization algorithms. *5th Conference on Evolutionary Algorithms and Global Optimization* (pp. 79–86). Warsaw: Warsaw University of Technology Press.

Hansen, N., & Ostermeyer, A. (2001). Completely derandomized self-adaptation in evolutionary strategies. *Evolutionary Computation, 9*(2), 159–195.

Hansen, N., Gemperle, F., Auger, A., & Koumoutsakos, P. (2006). When do heavy-tail distributions help? In T. Ph. Runarsson, H.-G Beyer, E. Burke, J. J. Merelo-Guervós, L.D. Whitley & X. Yao (Eds.), *Problem solving from nature (PPSN) IX* (Vol. 4193, pp. 62–71). Lecture Notes in Computer Science. Berlin: Springer.

Holland, J. H. (1992). *Adaptation in natural and artificial systems*. Cambridge: MIT Press.

Kappler, C. (1996). Are evolutionary algorithms improved by large mutation. In: H.-M. Voigt, W. Ebeling, I. Rechenberg & H.-P. Schwefel (Eds.), *Problem solving from nature (PPSN) IV* (Vol. 1141, pp. 388–397). Lecture Notes in Computer Science. Berlin: Springer.

Lee, C. Y., & Yao, X. (2004). Evolutionary programming using mutation based on the Lévy probability distribution. *IEEE Transactions on Evolutionary Computation, 8*(1), 1–13.

Michalewicz, Z. (1996). *Genetic algorithms + data structures = evolution programs*. Heidelberg: Springer.

Mitchel, M. (1996). *An Introduction to genetic algorithms*. Cambridge: MIT Press.

Nolan, J. P. (2007). *Stable distributions-models for heavy tailed data*. Boston: Birkhäuser.

Nolan, J. P., Panorska, A. K., & McCulloch, J. H. (2001). Estimation of stable spectral measures-stable non-Gaussian models in finance and econometrics. *Mathematical and Computer Modelling, 34*(9), 1113–1122.

Obuchowicz, A. (2003a). Multidimensional mutations in evolutionary algorithms based on real-valued representation. *International Journal of System Science*, *34*(7), 469–483.
Obuchowicz, A. (2003b). *Evolutionary algorithms in global optimization and dynamic system diagnosis*. Zielona Góra: Lubuskie Scientific Society.
Obuchowicz, A., & Prętki, P. (2004a). Evolutionary algorithms with α-stable mutations. In *IEEE 4th International Conference on Intelligent Systems Design and Application, Budapest, Hungary*, CD-ROM.
Obuchowicz, A., & Prętki, P. (2004b). Phenotypic evolution with mutation based on symmetric α-stable distributions. *International Journal on Applied Mathematics and Computer Science*, *14*(3), 289–316.
Obuchowicz, A., & Prętki, P. (2005). Isotropic symmetric α-stable mutations for evolutionary algorithms. In *IEEE congress on evolutionary computation* (pp. 404–410). Edinburgh, UK.
Obuchowicz, A., & Prętki, P. (2010). Evolutionary algorithms with stable mutational based on a discrete spectral measure. In L. Rutkowski, R. Scherer, R. Tadeusiewicz, L. A. Zadeh, & J. M. Zurada (Eds.), *Artificial intelligence and soft computing: Part II* (Vol. 6114, pp. 181–188). Lecture Notes on Artificial Intelligence. Berlin: Springer.
Obuchowicz, A. K., & Smołka, M. (2016). Application of α-stable mutation in hierarchic evolutionary inverse solver. *Journal on Computer Science*, *17*, 261–269.
Obuchowicz, A. K., Smołka, M., & Schaefer, R. (2015). Hierarchic genetic search with α-stable mutation. In A. I. Esparcia-Alcázar, & A. M. Mora, (Eds), *Applications of evolutionary computation* (Vol. 9028, pp. 143–154). Lecture Notes in Computer Science. Berlin: Springer.
Prętki P., & Obuchowicz A. (2006). Directional distributions and their application to evolutionary algorithms. In L. Rutkowski, R. Scherer, R. Tadeusiewicz, L. A. Zadeh, & J. M. Zurada (Eds.), *Artificial intelligence and soft computing* (Vol. 4029, pp. 440–449). Lecture Notes on Artificial Intelligence. Berlin: Springer.
Rechenberg, I. (1965). Cybernetic solution path of an experimental problem. In *Royal aircraft establishment*, Library Translation, *1122*, Hants: Farnborough.
Rudolph, G. (1997). Local convergence rates of simple evolutionary algorithms with Cauchy mutations. *IEEE Transactions on Evolutionary Computation*, *1*(4), 249–258.
Samorodnitsky, G., & Taqqu, M. S. (1994). *Stable non-Gaussian random processes*. New York: Chapman and Hall.
Schalkoff, R. J. (1990). *Artificial intelligence: An engineering approach*. New York: McGraw-Hill.
Schwefel, H.-P. (1995). *Evolution and optimum seeking*. New York: Wiley.
Trojanowski, K. (2008). *Practical metaheuristics*. Warsaw (in Polish): WIT Press.
Trojanowski, K. (2009). Properties of quantum particles in multi-swarm for dynamic optimization. *Fundamenta Informaticae*, *95*(2–3), 349–380.
Trojanowski, K., & Wierzchon, S. (2009). Immune-based algorithms for dynamic optimization. *Information Sciences*, *179*, 1495–1515.
Trojanowski K., Raciborski, M., & Kaczynski, P. (2013). Adaptive differential evolution with hybrid rules of perturbation for dynamic optimization. In K. Madani, A. Dourado, A. Rosa, & J. Filipe (Eds.), *Computational intelligence* (Vol. 465, pp. 69–83). Studies in Computational Intelligence. Berlin: Springer.
Yao, X., & Liu, Y. (1996). Fast evolutionary programming. *5th Annual Conference on Evolutionary Programming* (pp. 419–429). Cambridge: MIT Press.
Yao, X., & Liu, Y. (1997). Fast evolutionary strategies. Control. *Cybernetics*, *26*(3), 467–496.
Yao, X., & Liu, Y. (1999). Evolutionary programming made faster. *IEEE Transactions on Evolutionary Computation*, *3*(2), 82–102.
Zieliński, R. A., & Neumann, P. (1983). *Stochastic methods of the function minimum searching*. Berlin: Springer.

Chapter 2
Foundations of Evolutionary Algorithms

Evolutionary algorithms are a broad class of stochastic adaptation algorithms inspired by biological evolution—the process that allows populations of organisms to adapt to their surrounding environment. The concept of evolution was introduced in the 19th century by Charles Darwin and Johann Gregor Mendel and, complemented with further details, is still widely acknowledged as valid.

In 1859, Darwin published his theory of natural selection, or survival of the fittest. The idea behand it is as follows: not every organism can be kept, only those which can adapt and win the competition for food and shelter are able to survive. Almost at the same time (1865) Mendel published a short monograph about experiments with plant hybridization. He observed how traits of different parents are combined in a descendant by sexual reproduction. Darwinian evolutionary theory and Mendel's investigations of heredity in plants became foundations of evolutionary search methods.

The structure and properties of evolutionary algorithms are discussed in several books (Angeline and Kinnear 1996; Arabas 2001; Bäck 1995; Bäck et al. 1997; Dasgupta and Michalewicz 1997; Davis 1987; Fogel 1998; Galar 1990; Goldberg 1989; Holland 1992; Michalewicz 1996; Mitchel 1996; Schaefer 2007; Schwefel 1995). Papers concerned with evolutionary computation are published in many scientific journals. There are at least 20 international conferences closely connected with evolutionary methods. Due to the large number of available publications, it is impossible to present all of the different evolutionary algorithms and their components, whose authors tried to improve algorithm efficiency in the case of a given problem to be solved. A relatively wide glance at population methods applied to the global optimization problem can be found in the monographs by Arabas (2001), Bäck et al. (1997) and Schaefer (2007). In this chapter, the main components of evolutionary algorithms are recalled and their various basic forms are briefly discussed.

A. Obuchowicz, *Stable Mutations for Evolutionary Algorithms*,
Studies in Computational Intelligence 797,
https://doi.org/10.1007/978-3-030-01548-0_2

2.1 Basic Concepts

In nature, individuals in a population compete with one another for resources such as food, water, and shelter. Also, members of the same species often compete to attract a mate. The individuals which are most successful in surviving and attracting mates will have relatively larger numbers of descendants. Poorly performing individuals will produce few or even no descendant at all. This means that the information (genes), slightly mutated, from the highly adapted individuals will spread to an increasing number of individuals in each successive generation. In this way, species evolve to become more and more well-suited to their environment.

In order to describe a general outline of the evolutionary algorithm, let us introduce some useful concepts and notations (Atmar 1992; Fogel 1999; Schaefer 2007). An individual (a sample solution for some optimization or adaptation task) is represented by the genotype (an underlying genetic coding) $\boldsymbol{a} \in \mathcal{G}$, which is an element of some *genetic universum* $\mathcal{G}$. A population $P \in \mathcal{P}_\eta(\mathcal{G})$ is some multiset of η elements from $\mathcal{G}$. $\mathcal{P}_\eta(\mathcal{G})$ represent a set of all possible such multisets. The environment delivers quality information (fitness value) of the individual dependent on its phenotype (the manner of response contained in the behavior, physiology and morphology of the organism). The *fitness function*

$$\Phi : \mathcal{D} \to \mathbb{R} \tag{2.1}$$

is defined on a *phenotype space* $\mathcal{D}$. So, each individual can be viewed as a duality of its genotype and phenotype, and so some decoding function, *epigenesis*,

$$\xi : \mathcal{G} \to \mathcal{D}' \subset \mathcal{D}, \tag{2.2}$$

is needed.

At the beginning, an initial population $P(0)$ is generated (in a random or deterministic manner) and evaluated, i.e., the fitness function is calculated for each element of $P(0)$ (Table 2.1). After *initiation*, the randomized processes of *reproduction*, *recombination*, *mutation* and *succession* are iteratively repeated until a given termination criterion,

$$\iota : \mathcal{P}_\eta(\mathcal{G}) \to \{\text{true, false}\}, \tag{2.3}$$

is satisfied.

Reproduction,

$$s^p_{\theta_p} : \mathcal{P}_\eta(\mathcal{G}) \to \mathcal{P}_{\eta'}(\mathcal{G}), \tag{2.4}$$

called also *preselection*, is a randomized process (deterministic in some algorithms) of selection of η' parents from η individuals of the current population. This process is controlled by a set θ_p of parameters.

Table 2.1 General outline of an evolutionary algorithm

I. Initiation

A. Random generation of the first population
$P(0) = \{\boldsymbol{a}_k^0 \mid k = 1, 2, \ldots, \eta\}$

B. Evaluation
$P(0) \to \Phi(P(0)) = \{\phi_k^0 = \Phi(\xi(\boldsymbol{a}_k^0)) \mid k = 1, 2, \ldots, \eta\}$

C. $t = 0$

II. Repeat

A. Reproduction
$P'(t) = s_{\theta_p}^p(P(t)) = \{\boldsymbol{a}'^t_k \mid k = 1, 2, \ldots, \eta'\}$

B. Recombination
$P''(t) = r_{\theta_r}(P'(t)) = \{\boldsymbol{a}''^t_k \mid k = 1, 2, \ldots, \eta''\}$

C. Mutation
$P'''(t) = m_{\theta_m}(P''(t)) = \{\boldsymbol{a}'''^t_k \mid k = 1, 2, \ldots, \eta''\}$

D. Evaluation
$P'''(t) \to \Phi(P'''(t)) = \{\phi_k^t = \Phi(\xi(\boldsymbol{a}'''^t_k)) \mid k = 1, 2, \ldots, \eta''\}$

E. Succession
$P(t+1) = s_{\theta_n}^n(P(t) \cup P'''(t)) = \{\boldsymbol{a}_k^{t+1} \mid k = 1, 2, \ldots, \eta\}$

F. $t = t + 1$

Until $(\iota(P(t)) = \text{true})$

Recombination,

$$r_{\theta_r} : \mathscr{P}_{\eta'}(\mathscr{G}) \to \mathscr{P}_{\eta''}(\mathscr{G}), \tag{2.5}$$

is a mechanism (omitted in some realizations), controlled by additional parameters θ_r, which allows the mixing of parental information while passing it to their descendants.

Mutation

$$m_{\theta_m} : \mathscr{P}_{\eta''}(\mathscr{G}) \to \mathscr{P}_{\eta''}(\mathscr{G}), \tag{2.6}$$

introduces innovation into current descendants, θ_m is again a set of control parameters.

Succession,

$$s_{\theta_n}^n : \mathscr{P}_{\eta}(\mathscr{G}) \times \mathscr{P}_{\eta''}(\mathscr{G}) \to \mathscr{P}_{\eta}(\mathscr{G}), \tag{2.7}$$

also called *postselection*, is applied to choose a new generation of individuals from parents and descendants.

Considering reproduction $s_{\theta_p}^p$, recombination r_{θ_r}, mutation m_{θ_m} and succession $s_{\theta_n}^n$ operators, it is worth noting that they are random operators in the general case. For the purpose of analysis of these operators and their influence on an evolutionary process, target sets should be defined as some probabilistic space. But, if they are treated as elements of a particular algorithm, in which given realizations of the random variables are obtained as their result, we assume that the target sets are the Cartesian product of some number of genotype universes $\mathscr{G}$.

2.2 Standard Evolutionary Algorithms

Apart from similarities among various evolutionary computation techniques, there are also many differences. It is generally accepted that any evolutionary algorithm to solve a problem must have five basic components (Davis 1987; Michalewicz 1999):

- a representation of solutions to the problem,
- a way to create an initial population of solutions,
- an evaluation function, rating solutions in terms of their fitness,
- selection and variation operators that alter the composition of children during reproduction and mutation,
- values for the parameters (population size, probabilities of applying variation operators, etc.).

The variety of possible realizations of the above-mentioned components produces very large variants of evolutionary algorithms, which are usually specialized and dedicated to a particular optimization or adaptation task.

The duality of the genotype and the phenotype suggests two main approaches to simulated evolution dedicated to global optimization problems in $\mathbb{R}^n$: genotypic and phenotypic simulations (Fogel 1999). The former are focused on genetic structures. The candidate solutions are described as being analogous to chromosomes. The entire searching process is provided in the genotype space $\mathscr{G}$. However, in order to calculate an individual's fitness, its chromosome must be decoded to its phenotype. Two main streams of instances of such evolutionary algorithms can nowadays be identified:

- genetic algorithms (GAs) (De Jong 1975; Goldberg 1989; Grefenstette 1986; Holland 1975; Michalewicz 1996),
- genetic programming (GP) (Kinnear 1994; Koza 1992).

In phenotypic simulations, attention is focused on the behaviors of the candidate solutions in a population. All searching operations, i.e., selection, reproduction and mutation, are constructed in the phenotype space $\mathscr{D}$. This type of simulation characterizes a strong behavioral link between a parent and its descendant. Nowadays, there are two main streams of instances of 'phenotypic' evolutionary algorithms:

- evolutionary programming (EP) (Fogel et al. 1966, 1991; Fogel 1999; Yao and Liu 1999),
- evolutionary strategies (ESs) (Rechenberg 1965; Schwefel 1981).

Besides the above-mentioned algorithms, one more phenotypic model is of crucial importance in the research described in this book. This is the evolutionary search with soft selection (ESSS) algorithm (Galar 1985), which is a simplified version of the ES. The basic ideas behind GA, GP, EP and ES algorithms are presented below. The ESSS algorithm is the subject of the next section.

2.2.1 Genetic Algorithms

GAs are probably the best known evolutionary algorithms, receiving remarkable attention all over the world. The basic principles of GAs were first laid down rigorously by Holland (1975), and are well described in many texts (e.g. Bäck and Schwefel 1993; Beasley et al. 1993a, b Dasgupta and Michalewicz 1997; Davis 1987, 1991; Goldberg 1989; Grefenstette 1986, 1990, 1990; Michalewicz 1996).

The previously proposed forms of GAs (De Jong 1975; Holland 1975) operate on binary strings of fixed length l, i.e., the genotype space $\mathscr{G}$ is an l-dimensional Hamming cube $\mathscr{G} = \{0, 1\}^l$. A GA with the binary representation of the individual is a natural technique of solving discrete problems, especially in the case of finite cardinality of possible solutions. Such a problem can be transformed to a pseudo-Boolean fitness function, where the GA can be used directly. In the case of continuous domains of optimization problems, the function $\zeta : \mathscr{D} \to \mathscr{G}$ that encodes the variables of the given problem into a bit string (the so-called chromosome) is needed. The encoding function ζ is non-invertible and there does not exist the inverse function ζ^{-1}. A decoding function $\xi : \mathscr{G} \to \mathscr{D}_l \subset \mathscr{D}$ generates only 2^l representatives of solutions. This is a strong limitation of GAs. Moreover, it is worth noting that, taking into account different topologies of genotype and phenotype spaces, the transformation $\xi : \mathscr{D} \to \mathscr{G}$ should fulfil additional, very difficult to obtain, conditions, which prevent occurence of a new local optimum in $\mathscr{G}$ in comparison to $\mathscr{D}$ (Arabas 2001).

There are many known GA versions in the literature. In this chapter the version described by Arabas (2001) is presented. The parent selection s^p is carried out by the so-called *proportional method* (*roulette method*):

$$s^p\big(P(t)\big) = \big\{\boldsymbol{a}_{h_1}, \boldsymbol{a}_{h_2}, \ldots, \boldsymbol{a}_{h_\eta}\big\} : \quad h_k = \min\left\{h : \frac{\sum_{l=1}^{h} \phi_l^t}{\sum_{l=1}^{\eta} \phi_l^t} > \chi_k\right\}, \tag{2.8}$$

where $\{\chi_k = U(0, 1) \mid k = 1, 2, \ldots, \eta\}$ are uniformly distributed, independent random numbers from the interval $[0, 1)$. In this type of selection, the probability that a given chromosome will be chosen as a parent is proportional to its fitness. Because sampling is carried out with returns, it can be expected that well-fitted individuals insert a few of their copies in the temporary population $P'(t)$.

After the reproduction, chromosomes from $P'(t)$ are recombined using the so-called *one-point crossover* operation. Chromosomes from $P'(t)$ are joined into pairs. The decision that a given pair will be recombined is made with the given probability θ_r. If the decision is positive, an ith position in the chromosome is randomly chosen and the information from the position $(i + 1)$ to the end of chromosomes is exchanged in the pair:

$$\left\{\begin{matrix} (a_1, a_2, \ldots, a_l) \\ (b_1, b_2, \ldots, b_l) \end{matrix}\right\} \to \left\{\begin{matrix} (a_1, \ldots, a_i, b_{i+1}, \ldots, b_l) \\ (b_1, \ldots, b_i, a_{i+1}, \ldots, a_l) \end{matrix}\right\}.$$

The newly obtained temporary population $P''(t)$ is mutated. The individuals' mutation m_{θ_m} is performed separately for each bit in a chromosome. The bit value is changed to the opposite one with the given probability θ_m. The obtained population is the population of a new generation.

Historically, the first attempt at a formal description of the asymptotic characteristics of the GA was made by Holland (1975). The combined effect of selection, crossover and mutation gives the so-called reproductive schema growth equation (Schaefer 2007):

$$E[\eta(S, t+1)] \geq \eta(S, t)\frac{\Phi(S, t)}{\bar{\Phi}(t)}\left(1 - \theta_r\frac{\delta(S)}{l-1}\right)(1-\theta_m)^{o(S)}, \tag{2.9}$$

where S is a schema defined as a string of l symbols from an alphabet $\Sigma = \{0, 1, \star\}$, each schema represents all strings which match it on all positions other than '$\star$', $\eta(S, t)$ denotes the number of strings in a population at the time t matched by schema S, $E[\cdot]$ is a symbol of an expectation value, $\delta(S)$ is the defining length of the schema S (the distance between the first and the last fixed string position), $o(S)$ denotes the order of the schema S (the number of 0 and 1 positions presented in the schema), $\Phi(S, t)$ is defined as the average fitness of the all strings in the population at the time t matched by the schema S, and $\bar{\Phi}(t)$ is the average fitness taken over all individuals in the population at the time t.

Equation (2.9) tells us about the expected number of strings matching a schema S in the next generation as a function of the current number of strings matching the schema, the relative fitness of the schema, as well as its defining length and order. It is clear that above-average schemata with a short defining length and a low order would still be sampled at exponentially increased rates.

The above approach, often criticized (see Schaefer 2007; Grefenstette 1993), can be treated as an attempt at the evaluation of numerical improvement of population quality (Whitley 1994). Vose (1999) proves, under some additional conditions, that a Markov process, which models genetic algorithm processing, is ergodic. This fact implies asymptotic correctness in the probabilistic sense and the asymptotic guarantee of success (Schaefer 2007).

2.2.2 Genetic Programming

Many trends in SGA development are connected with the change of an individual representation. One of them deserves particular attention: each individual is a tree (Koza 1992). This small change in the GA gives evolutionary techniques the possibility of solving problems which have not been tackled yet. This type of GA is called genetic programming (GP).

Two sets are needed to be defined before GP starts: the set of terms $\mathscr{T}$ and the set of operators $\mathscr{F}$. In the initiation step, the population of trees is randomly chosen. For each tree, leaves are chosen from the set $\mathscr{T}$ and other nodes are chosen from the set $\mathscr{F}$. Depending on definitions $\mathscr{T}$ and $\mathscr{F}$, a tree can represent a polyadic function, a

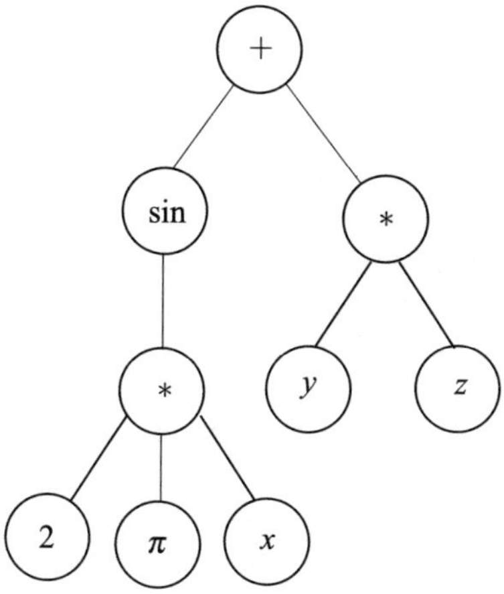

Fig. 2.1 Sample of a tree which represents the function $f(x, y, z) = yz + \sin(2\pi x)$

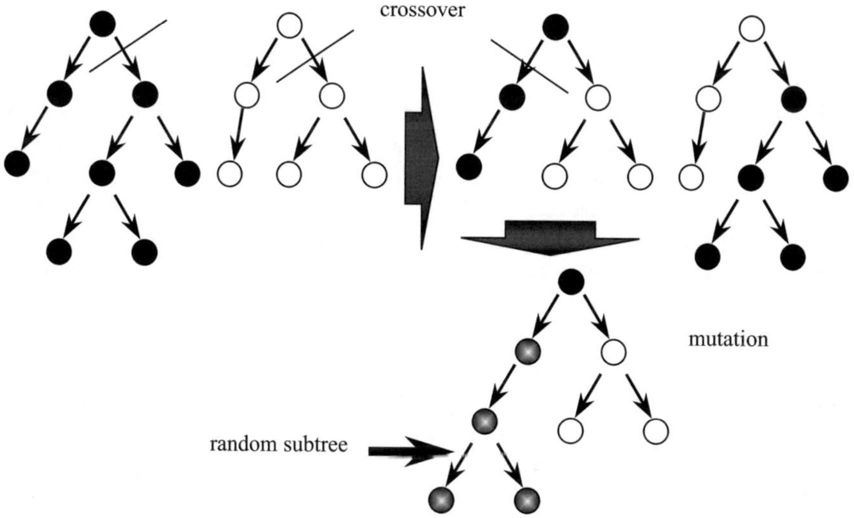

Fig. 2.2 Genetic operators for GP

logical sentence or a part of a programming code in a given programming language. Figure 2.1 presents a sample tree for $\mathcal{T} = \{x, y, z, 2, \pi\}$ and $\mathcal{F} = \{*, +, -, sin\}$.

The new type of individual representation needs new definitions of crossover and mutation operators, both explained in Fig. 2.2.

2.2.3 Evolutionary Programming

Evolutionary programming resides in the 'phenotypic' category of simulations. It was devised by Fogel et al. (1966) in the mid-1960s for the evolution of finite state machines in order to solve prediction tasks. The environment was described as a sequence of symbols (from a finite alphabet), and a finite state machine had to create a new symbol. The output symbol had to maximize some profit function, which was a measure of prediction accuracy. There are no reproduction or recombination operators. Each machine of the current population generates a descendant by random mutation. There are five possible modes of random mutation that naturally result from the description of the finite state machine: changing an output symbol, changing a state transition, adding a state, deleting a state, or changing the initial state. Mutation is chosen with respect to a uniform distribution. The best half number of parents and descendant are selected to survive.

EP was extended by Fogel (1991, 1992) to work on real-valued object variables based on normally distributed mutations. This algorithm was called meta-EP (Fogel et al. 1991) or classical EP (CEP) (Yao and Liu 1999). The description shown in Table 2.2 is based on the work of Bäck and Schwefel (1993), as well as Yao and Liu (1999).

In meta-EP, an individual is represented by a pair $\boldsymbol{a} = (\boldsymbol{x}, \boldsymbol{\sigma})$, where $\boldsymbol{x} \in \mathbb{R}^n$ is a real-valued phenotype while $\boldsymbol{\sigma} \in \mathbb{R}^n_+$ is a self-adapted standard deviation vector for Gaussian mutation. For initialization, EP assumes bounded initial domains $\Omega_x = \prod_{i=1}^n [u_i, v_i] \subset \mathbb{R}^n$ and $\Omega_\sigma = \prod_{i=1}^n [0, c] \subset \mathbb{R}^n_+$, with $u_i < v_i$ and $c > 0$. However, the search domain is extended to $\mathbb{R}^n \times \mathbb{R}^n_+$ during algorithm processing. As a mutation operator, Gaussian mutation with a standard deviation vector ascribed to an individual is used. All elements in the current population are mutated. Individuals from both parent and descendant populations participate in the process of selecting a new generation. For each individual $\boldsymbol{a}_k$, q individuals are chosen at random from $P(t) \cup P'(t)$ and compared to $\boldsymbol{a}_k$ with respect to their fitness values. w_k is the number of q individuals worse than $\boldsymbol{a}_k$. η individuals having the highest score w_k are selected from 2η parents and descendants to form a new population $P(t+1)$.

The analysis of the classical EP algorithm (Fogel 1992) gives a proof of the global convergence with probability one for the resulting algorithm, and the result is derived from defining a Markov chain over the discrete state space that is obtained from a reduction of the abstract search space $\mathbb{R}^n$ to the finite set of numbers representable on a digital computer.

2.2.4 Evolutionary Strategies

The other well-known 'phenotypical' algorithms are evolutionary strategies, which were introduced in the mid-1960s by Rechenberg (1965) and further developed by Schwefel (1981). The description of the ES presented in this subsection is based on the article by Bäck and Schwefel (1993). The general form of the ES relies on the

Table 2.2 Outline of the EP algorithm

I. Initialize

A. Random generation

$P(0) = \{a_k^0 = (x_k^0, \sigma_k(0)) \mid k = 1, 2, \ldots, \eta\}$

$x_k^0 = RANDOM(\Omega_x), \sigma_k^0 = RANDOM(\Omega_\sigma),$

$\Omega_x \subset \mathbb{R}^n, \Omega_\sigma \subset \mathbb{R}_+^n$

B. Evaluation

$P(0) \to \Phi(P(0)) = \{\phi_k^0 = \Phi(x_k^0) \mid k = 1, 2, \ldots, \eta\}$

C. $t = 1$

II. Repeat

A. Mutation

$P'(t) = m_{\tau,\tau'}(P(t)) = \{a'^t_k \mid k = 1, 2, \ldots, \eta'\}$

$x'^t_{ki} = x^t_{ki} + \sigma^t_{ki} N_i(0, 1), \sigma'^t_{ki} = \sigma^t_{ki} \exp(\tau' N(0, 1) + \tau N_i(0, 1)),$

$i = 1, 2, \ldots, n,$

where $N(0, 1)$ denotes a normally distributed one-dimensional random number with mean zero and standard deviation one, $N_i(0, 1)$ indicates that the random number is generated anew for each component i

B. Evaluation

$P'(t) \to \Phi(P'(t)) = \{\phi'^t_k = \Phi(x'^t_k) \mid k = 1, 2, \ldots, \eta\}$

C. Succession

$P(t+1) = s^n_{\theta_n}(P(t) \cup P'(t)) = \{a_k^{t+1} \mid k = 1, 2, \ldots, \eta\}$

$\forall a_k^t \in P(t) \cup P'(t),$

$a_k^t \to \{a_{kl}^t = RANDOM(P(t) \cup P'(t)) \mid l = 1, 2, \ldots, q\},$

$w_k^t = \sum_{l=1}^q \theta(\Phi(x_k^t) - \Phi(x_{kl}^t)), \theta(\alpha) = \begin{cases} 0 & \text{for } \alpha < 0 \\ 1 & \text{for } \alpha \geq 0 \end{cases},$

$P(t+1) \leftarrow \eta$ individuals with the best scorew_k^t

D. $t = t + 1$

Until $(\iota(P(t)) = \text{true})$

individual representation in the form of a pair $\boldsymbol{a} = (\boldsymbol{x}, \mathbb{C})$, where $\boldsymbol{x} \in \mathbb{R}^n$ is a point in a searching space, and the fitness value of the individual $\boldsymbol{a}$ is calculated directly from the objective function: $\Phi(\boldsymbol{a}) = f(\boldsymbol{x})$. $\mathbb{C}$ is the covariance matrix for the general n-dimensional normal distribution $N(\mathbf{0}, \mathbb{C})$, having the probability density function

$$p(z) = \sqrt{\frac{1}{(2\pi)^n \det(\mathbb{C})}} \exp\left(-\frac{1}{2} z^T \mathbb{C}^{-1} z\right), \tag{2.10}$$

where $z \in \mathbb{R}^n$. To assure positive-definiteness of $\mathbb{C}$, it is described by two vectors: the vector of standard deviations $\boldsymbol{\sigma}$ $(c_{ii} = \sigma_i^2)$ and that of rotation angles $\boldsymbol{\alpha}(c_{ij} = \frac{1}{2}(\sigma_i^2 - \sigma_j^2) \tan 2\alpha_{ij})$. Thus, $\boldsymbol{a} = (\boldsymbol{x}, \boldsymbol{\sigma}, \boldsymbol{\alpha})$ is used to denote a complete individual.

There is no separate operation of selection of parents in ESs; this selection is strongly connected with the recombination mechanism. Different recombination mechanisms can be used in ESs to create λ new individuals. Recombination rules of determining an individual $\boldsymbol{a}' = (\boldsymbol{x}', \boldsymbol{\sigma}', \boldsymbol{\alpha}')$ have the following form:

$$a_i' = \begin{cases} a_{p,i} & \text{without recombination,} \\ a_{p,i} \text{ or } a_{s,i} & \text{discrete recombination,} \\ a_{p,i} + \chi(a_{s,i} - a_{p,i}) & \text{intermediate recombination,} \\ a_{p_i,i} \text{ or } a_{s_i,i} & \text{global discrete recombination,} \\ a_{p_i,i} + \chi_i(a_{s_i,i} - a_{p_i,i}) & \text{global intermediate recombination,} \end{cases} \tag{2.11}$$

where indices p and s denote two parent individuals selected at random from $P(t)$, and $\chi \in [0, 1]$ is uniform random variable. For global variants, for each component of $\boldsymbol{a}$, the parents p_i, s_i as well as χ_i are determined anew. Empirically, discrete recombination of object variables and intermediate recombination of strategy parameters have been observed to give the best results (Bäck and Schwefel 1993).

Each recombined individual $\boldsymbol{a}'$ is subject to mutation. Firstly, strategy parameters are mutated, and then new object variables are calculated using new standard deviations and rotation angles:

$$\begin{aligned} \boldsymbol{\sigma}'' &= \{\sigma_i' \exp(\tau' N(0,1) + \tau N_i(0,1)) \mid i = 1, 2, \ldots, n\}, \\ \boldsymbol{\alpha}'' &= \{\alpha_j' + \beta N_j(0,1) \mid j = 1, 2, \ldots, n(n-1)/2\}, \\ \boldsymbol{x}'' &= \boldsymbol{x}' + \mathbf{N}(\mathbf{0}, \boldsymbol{\sigma}'', \boldsymbol{\alpha}''), \end{aligned} \tag{2.12}$$

where factors τ', τ and β are rather robust exogenous parameters, which are suggested to be set as follows: $\tau' \approx 1/\sqrt{2\sqrt{n}}$, $\tau \approx 1/\sqrt{2n}$ and $\beta \approx 0.0873$ (5° in radians).

Selection in ESs is completely deterministic. There exist two possible strategies:

- $(\mu + \lambda)$-ES: selecting μ best individuals out of the union of μ parents and λ descendants,
- (μ, λ)-ES: selecting μ best individuals out of the set of λ descendants ($\lambda > \mu$).

Although $(\mu + \lambda)$-ES is elitist and guarantees monotonously improving performance, the effectiveness of global optimum searching is worse than in the case of (μ, λ)-ES; therefore, the latter is recommended nowadays.

Under some restrictions, it is possible to prove the convergence theorem for evolutionary strategies (Bäck et al. 1991). Let the covariance matrix $\mathbb{C}$ be reduced to the standard deviation vector which has all components identical, i.e., $\boldsymbol{\sigma} = \{\sigma, \ldots, \sigma\}$ and $\sigma > 0$, and remains unchanged during the process. If the optimization problem with $\Phi_{opt} > -\infty$ (minimization) or $\Phi_{opt} < \infty$ (maximization) is regular, then the evolutionary process converges to the global optimum in an infinite limit of time with probability one.

Table 2.3 Outline of the ESSS algorithm

Input

η: population size;

$t_{\max}$: maximum number of iterations;

σ: standard deviation of mutation;

$\Phi : \mathbb{R}^n \to \mathbb{R}_+$: non-negative fitness function;

n: number of features;

$\boldsymbol{x}_0^0$: initial point

1. Initialize

(a) $P(0) = \{\boldsymbol{x}_1^0, \boldsymbol{x}_2^0, \ldots, \boldsymbol{x}_\eta^0\} : (\boldsymbol{x}_k^0)_i = (\boldsymbol{x}_0^0)_i + N(0, \sigma^2)$

$i = 1, 2, \ldots, n; \quad k = 1, 2, \ldots, \eta$

(b) $\phi_0^0 = \Phi(\boldsymbol{x}_0^0)$

2. Repeat

(a) Evaluation

$\Phi(P(t)) = \{\phi_1^t, \phi_2^t, \ldots, \phi_\eta^t\}$ where $q_k^t = \Phi(\boldsymbol{x}_k^t), \quad k = 1, 2, \ldots, \eta$

(b) Selection

$$\{h_1, h_2, \ldots, h_\eta\} \text{ where } h_k = \min\left\{h : \frac{\sum_{l=1}^h \phi_l^t}{\sum_{l=1}^\eta \phi_l^t} > \zeta_k\right\}$$

i $\{\zeta_k\}_{k=1}^\eta$ are random numbers uniformly distributed in $[0, 1)$

(c) Mutation

$P(t) \to P(t+1)$;

$(\boldsymbol{x}_k^{t+1})_i = (\boldsymbol{x}_{h_k}^t)_i + N(0, \sigma^2), \quad i = 1, 2, \ldots, n; \quad k = 1, 2, \ldots, \eta$

Until $(\iota(P(t)) = \text{true})$

2.2.5 Evolutionary Search with Soft Selection

The ESSS algorithm was introduced by Galar (1989) and is based on probably the simplest model of Darwinian phenotypical evolution (Galar 1985). This selection-mutation process is executed a multi-dimensional real space, on which the fitness function is defined (Table 2.3).

At the beginning, the first population $P(0)$ composed of η individuals is randomly selected. If the ESSS algorithm is chosen to solve the unconstrained global optimization problem, the concept stating that an initial population has to be uniformly distributed in the search space has no sense. One of the possible and rational solutions is to create an initial population by adding η times a normally distributed random vector to a given initial point $\boldsymbol{x}_0^0 \in \mathbb{R}^n$. The fitness $\phi_k^0 = \Phi(\boldsymbol{x}_k^0)$ is calculated for each element $\boldsymbol{x}_k^0$ of the population ($k = 1, 2, \ldots, \eta$). The searching process consists in generating a sequence of η-element populations. A new population $P(t+1)$ is created based only on the previous population $P(t)$. In order to generate a new element $\boldsymbol{x}_k^{t+1}$, a *parent* element is selected and mutated. Both selection and mutation are random processes. Each element $\boldsymbol{x}_k^t$ can be chosen as a parent with a probability proportional to its fitness ϕ_k^t (the *roulette method* (2.8)). A new element $\boldsymbol{x}_k^{t+1}$ is obtained by adding a normally distributed random value to each entry of the selected parent:

$$\left(x_k^{t+1}\right)_i = \left(x_{h_k}^t\right)_i + N(0, \sigma^2) \quad i = 1, \ldots, n, \tag{2.13}$$

where the standard deviation σ is a parameter to be elected. It is important to note that there is no recombination (crossover) operator in ESSS. However, the recombination operator is biologically motivated (Mendel's experiments) and is of great importance for EAs based on the genotypic representation of individuals. In the case of phenotype simulations of evolution, which are based on the floating point representation of individuals, mutation seems to be the crucial operator of the evolutionary process (Fogel 1995; Galar 1989).

The first attempt at ESSS convergence analysis was presented by Karcz-Dulęba (1997, 2001a), who considered dynamics of infinite populations in a landscape of unimodal and bimodal fitness functions. Galar and Karcz-Dulęba (1994) propose to consider evolutionary dynamics in the state space of the population. The population state space is $n\eta$-dimensional. Because evolutionary dynamics are independent of the elements' sequence in the population, the population state space does not cover all of the $\mathbb{R}^{n\eta}$ space but only some convex, compact and multi-lateral subspace of $\mathbb{R}^{n\eta}$. Analytical results for the population of two elements, obtained by using the population state space description, are presented by Karcz-Dulęba (2001, 2004).

Using proportional selection (the roulette method) as a reproduction operator is connected with some disadvantages. If the differences between fitness function values of particular individuals in a given population are much lower than their absolute values, then the selection pressure is almost equal to the uniform distribution. This means that the evolutionary process is independent of the fitness function. One of the possible effective solutions of this problem is proposed by Obuchowicz (2003c). Let $f : \mathbb{R}^n \to \mathbb{R}$ be a minimized objective function. Then the fitness function is defined as follows:

$$\Phi(x) = f_{\max}^t - f(x) + \frac{1}{\eta^2}, \tag{2.14}$$

where $f_{\max}^t = \max f\left(x_k^t \mid k = 1, \ldots, \eta\right)$ is the maximum value of the function f obtained in the current population. The last factor in (2.14) defines the selection probability of the worst individual in the current population. This probability should be respectively small, but not equal to zero. The basic upper limitation of this probability is equal to $1/\eta$, which is the selection probability of each individual in the case of reproduction based on the uniform distribution. The fitness function defined in (2.14) is non-negative and its relative values for different elements of the current population allow the reproduction to be effective.

Another way to avoid the above-mentioned disadvantage is the application of one more selection method—tournament selection. First, the tournament group P_G is composed of η_G uniformly chosen individuals of the current population. The best individual of P_G is chosen. Such a process is repeated η-times. The parameter η_G controls the selection pressure from the uniform distribution ($\eta_G = 1$) to the deterministic case ($\eta_G = \eta$). The ESSS algorithm with tournament selection will be denoted as ESTS in this book.

2.3 Summary

The evolutionary algorithm is distinguished by two main characteristics. Unlike other classes of optimization algorithms, the EA operates on the population of individuals. In this way, the knowledge about the environment is discovered simultaneously by many individuals, verifies information inherited from ancestors and is passed down from generation to generation. Species acquire their individual characteristics due to the survival of well-fitted ones, which is seemingly a blind mechanism where only individuals well adapted to presence can survive and procreate. However, nature does not select only the best individuals to procreate—sometimes even a weakly adapted one has a possibility of creating an descendant which can possess a feature without parallel in the population. This is the second evolution characteristic, called *soft selection*. If we give up hard selection and use soft one instead, assuming that weakly adapted points (in the sense of the values of the objective function) can be selected to create descendants, the possibility of finding the global optimum increases.

References

Angeline, P., & Kinnear, K. E. (1996). *Advances in genetic programming*. Cambridge: MIT Press.

Arabas, J. (2001). *Lectures on evolutionary algorithms*. Warsaw: WNT. (in Polish).

Atmar, W. (1992). On the rules and nature of simulated evolutionary programming. In D. B. Fogel & W. Atmar (Eds.), *1st Annual Conference on Evolutionary Programming* (pp. 17–26). Jolla: Evolutionary Programming Society.

Bäck, T. (1995). *Evolutionary algorithms in theory and practice*. Oxford: Oxford University Press.

Bäck, T., & Schwefel, H.-P. (1993). An overview of evolutionary algorithms for parameter optimization. *Evolutionary Computation*, *1*(1), 1–23.

Bäck, T., Fogel, D. B., & Michalewicz, Z. (Eds.). (1997). *Handbook of evolutionary computation*. New York: Institute of Physics Publishing and Oxford University Press.

Bäck, T., Hoffmeister, F., & Schwefel, H.-P. (1991). A survey of evolution strategies. In R. Belew & L. Booker (Eds.), *4th International Conference on Genetic Algorithms* (pp. 2–9). Los Altos: Morgan Kauffmann Publishers.

Beasley, D., Bull, D. R., & Martin, R. R. (1993a). An overview of genetic algorithms. Part 1: Fundamentals. University. *Computing*, *15*(2), 58–69.

Beasley, D., Bull, D. R., & Martin, R. R. (1993b). An overview of genetic algorithms. Part 2: Research topics. University. *Computing*, *15*(4), 170–181.

Dasgupta, D., & Michalewicz, Z. (Eds.). (1997). *Evolutionary algorithms for engineering applications*. Heidelberg: Springer.

Davis, L. (Ed.). (1987). *Genetic algorithms and simulated annealing*. San Francisco: Morgan Kaufmann.

De Jong, K. (1975). *An analysis of the behaviour of a class of genetic adaptive systems*. Ph.D. thesis, University of Michigan, Ann Arbor.

Fogel, D. B. (1992). An analysis of evolutionary programming. In*1st annual conference on genetic programming* (pp. 43–51). Jolla: Evolutionary Programming Society.

Fogel, D. B. (1995). *Evolutionary computation: toward a new philosophy of machine intelligence*. New York: IEEE Press.

Fogel, D. B. (1998). *Evolutionary computation: the fossil record*. NY: IEEE Press.

Fogel, D. B. (1999). An overview of evolutionary programming. In L. D. Davis, K. De Jong, M. D. Vose, & L. D. Whitley (Eds.), *Evolutionary algorithms* (pp. 89–109). Heidelberg: Springer.

Fogel, D. B., Fogel, L. J., & Atmar, J. W. (1991). Meta-evolutionary programming. *25th Asilomar Conference on Signals, Systems, and Computers* (pp. 540–545). San Jose: Maple Press.
Fogel, L. J., Owens, A. J., & Walsh, M. J. (1966). *Artificial intelligence through simulated evolution*. New York: Wiley.
Galar, R. (1985). Handicapped individua in evolutionary processes. *Biological Cybernetics* (vol. 51, pp. 1–9).
Galar, R. (1989). Evolutionary search with soft selection. *Biological cybernetics* (vol. 60, pp. 357–364).
Galar, R. (1990). *Soft Selection in Random Global Adaptation in R^n*. Wrocław (in Polish): A Biocybernetic Model of Development. - Technical University of Wrocław Press.
Galar, R., & Karcz-Dulęba, I. (1994). The evolution of two: An example of space of states approach. *3rd Annual Conference on Evolutionary Programming* (pp. 261–268). San Diego: World Scientific.
Goldberg, D.E. (1989). *Genetic algorithms in search, optimization and machine learning*. Reading: Addison-Wesley.
Grefenstette, J. J. (1986). Optimization od control parameters for genetic algorithms. *IEEE Transactions on System, Man and Cybernetics*, *16*(1), 122–128.
Grefenstette, J. J. (1990). Genetic algorithms and their applications. In A. Kent & J. G. Williams (Eds.), *Encyclopedia of computer Science and Technology* (pp. 139–152). New York: Marcel Dekker.
Grefenstette, J. J. (1993). Deception considerable harmful. *Foundations of Genetic Algorithms*, *2*, 75–91.
Holland, J. H. (1975). *Adaptation in natural and artificial systems*. Ann Arbor: The University of Michigan Press.
Holland, J. H. (1992). *Adaptation in natural and artificial systems*. Cambridge: MIT Press.
Karcz-Dulęba, I. (1997). Some convergence aspects of evolutionary search with soft selection method. *2nd Conference on Evolutionary Algorithms and Global Optimization* (pp. 113–120). Warsaw: Warsaw University of Technology Press.
Karcz-Dulęba, I. (2001a). Dynamics of infinite populations envolving in a landscape of uni- and bimodal fitness functions. *IEEE Transactions on Evolutionary Computation*, *5*(4), 398–409.
Karcz-Dulęba, I. (2001). Evolution of two-element population in the space of population states: Equilibrium states for assymetrical fitness functions. *5th Conference on Evolutionary Algorithms and Global Optimization* (pp. 106–113). Warsaw: Warsaw University of Technology Press.
Karcz-Dulęba, I. (2004). Time to convergence of evolution in the space of population states. *International Journal Applied Mathematics and Computer Science*, *14*(3), 279–287.
Kinnear, J. R. (Ed.). (1994). *Advances in genetic programming*. Cambridge: The MIT Press.
Koza, J. R. (1992). *Genetic programming: On the programming of computers by means of natural selection*. Cambridge: The MIT Press.
Michalewicz, Z. (1996). *Genetic algorithms + data structures = evolution programs*. Heidelberg: Springer.
Michalewicz, Z. (1999). The significance of the evaluation function in evolutionary algorithms. In L. D. Davis, K. De Jong, M. D. Vose, & L. D. Whitley (Eds.), *Evolutionary algorithms* (pp. 151–166). Heidelberg: Springer.
Mitchel, M. (1996). *An introduction to genetic algorithms*. Cambridge: MIT Press.
Obuchowicz, A. (2003c). Population in an ecological niche: Simulation of natural exploration. *Bulletin of the Polish Academy of Sciences: Technical Sciences*, *51*(1), 59–104.
Rechenberg, I. (1965). Cybernetic solution path of an experimental problem. *Royal aircraft establishment*, library translate 1122. Hants: Farnborough.
Schaefer, R. (2007). *Foundation of global genetic optimization*. Heidelberg: Springer.
Schwefel, H.-P. (1981). *Numerical optimization of computer models*. Chichester: Wiley.
Schwefel, H.-P. (1995). *Evolution and optimum seeking*. New York: Wiley.
Vose, M. D. (1999). *The simple genetic algorithm*. Cambridge: MIT Press.
Whitley, D. (1994). A genetic algorithm tutorial. *Statistics and computing* (vol. 4, pp. 65–85).
Yao, X., & Liu, Y. (1999). Evolutionary programming made faster. *IEEE Transactions on Evolutionary Computation*, *3*(2), 82–102.

Chapter 3
Stable Distributions

Stable distributions are defined in this chapter and their main characteristics are presented. The attention is focused on the properties which are directly useful for the analysis of stable distribution applicability to evolutionary processes. At the beginning, the definition of the stable random variable, as well as its characteristic function, is given. A series of theorems which describe the properties of the stable random value is presented, including the generalized central limit theorem, momentum existence and the formula of the sum of stable random variables. Procedures for stable pseudo-random values generation are shown. Next, the main properties of stable vectors are discussed. Non-isotropic random vectors, isotropic random vectors, and random vectors based on α-stable generators are considered.

3.1 Stable Random Value

3.1.1 Definition of Stable Distributions

If a probabilistic model of a given process is built, the term of a random value is used. Usually, this random value, as well as a probabilistic space of this process, is not directly defined. The model description is most often limited to a cumulative distribution function, a probability density function or a characteristic function (the Fourier transform of the probability density function).

The theory of stable distributions is of special importance in phenomenon modeling. The first works (P. Lévy, A.J. Khinchin) on this theory were publicized in the 1920 and 1930s. Stable distributions can be defined in four equivalent ways (Feller 1971; Samorodnitsky and Taqqu 1994; Nolan 2007). The basic definition takes into account the fact that the family of stable distributions forms the group with the sum operation of individual random variables (Samorodnitsky and Taqqu 1994).

A. Obuchowicz, *Stable Mutations for Evolutionary Algorithms*,
Studies in Computational Intelligence 797,
https://doi.org/10.1007/978-3-030-01548-0_3

Definition 3.1 A random variable X has a stable distribution if

$$\forall a, b > 0 \;\; \exists c > 0 \;\; \exists d \in \mathbb{R} : \quad aX_1 + bX_2 \stackrel{d}{=} cX + d, \tag{3.1}$$

where random variables X, X_1, X_2 have the same distribution and $\stackrel{d}{=}$ denotes equality in distribution.

It is easy to note that a degenerated random variable X, i.e., concentrated at some point x_0 ($P(X = x_0) = 1$), is always stable. This case is not interesting, so we will consider later only the non-degenerated one. If $d = 0$ in Definition 3.1, then we call the random variable strictly stable.

The theorem presented below describes a very important property of stable random variables (Feller 1971; Samorodnitsky and Taqqu 1994).

Theorem 3.1 *For each stable random variable X, there exists a number $\alpha \in (0, 2]$ such that c in Definition 3.1 fulfills the following equation:*

$$c^\alpha = a^\alpha + b^\alpha. \tag{3.2}$$

The proof can be found in Chapter VI.1 of the book by Feller (1971). The parameter α is called the *stability index* or *characteristic exponent*, and the random variable X with the stability index α is called the α-stable random variable.

The property described in the theorem presented above is extended by the following theorem (the proof is given by Feller (1971, Theorem VI.1.1)).

Theorem 3.2 *A random variable X has a stable distribution if*

$$\forall n \geqslant 2 \;\; \exists \alpha \in (0, 2] \;\; \exists d_n \in \mathbb{R} : \quad X_1 + X_2 + \cdots + X_n \stackrel{d}{=} n^{1/\alpha} X + d_n, \tag{3.3}$$

where $X_1, X_2, \ldots, X_n$ are independent copies of X.

Research works of Augustin Cauchy, George Polya, Paul Lévy and Aleksandr Khinchin allow us to formulate a precise description of the family of stable distributions which meet the Definition 3.1.

Theorem 3.3 *A random variable X has a stable distribution if there exist parameters $\alpha \in (0, 2]$, $\sigma \geq 0$, $\beta \in [-1, 1]$ and $\mu \in \mathbb{R}$ such that the characteristic function of X has the form*

$$\varphi(k) = \begin{cases} \exp\left(-\sigma^\alpha |k|^\alpha \left\{1 - i\beta \operatorname{sign}(k) \tan\left(-\frac{\pi\alpha}{2}\right)\right\} + i\mu k\right), & \alpha \neq 1, \\ \exp\left(-\sigma |k| \left\{1 + i\frac{2}{\pi}\beta \operatorname{sign}(k) \ln |k|\right\} + i\mu k\right), & \alpha = 1, \end{cases} \tag{3.4}$$

where α is called the stability index, β is the skew parameter, σ is the scale parameter, μ is the localization parameter, and

$$\operatorname{sign}(x) \begin{cases} 1 \text{ if } x > 0, \\ 0 \text{ if } x = 0, \\ -1 \text{ if } x < 0. \end{cases} \tag{3.5}$$

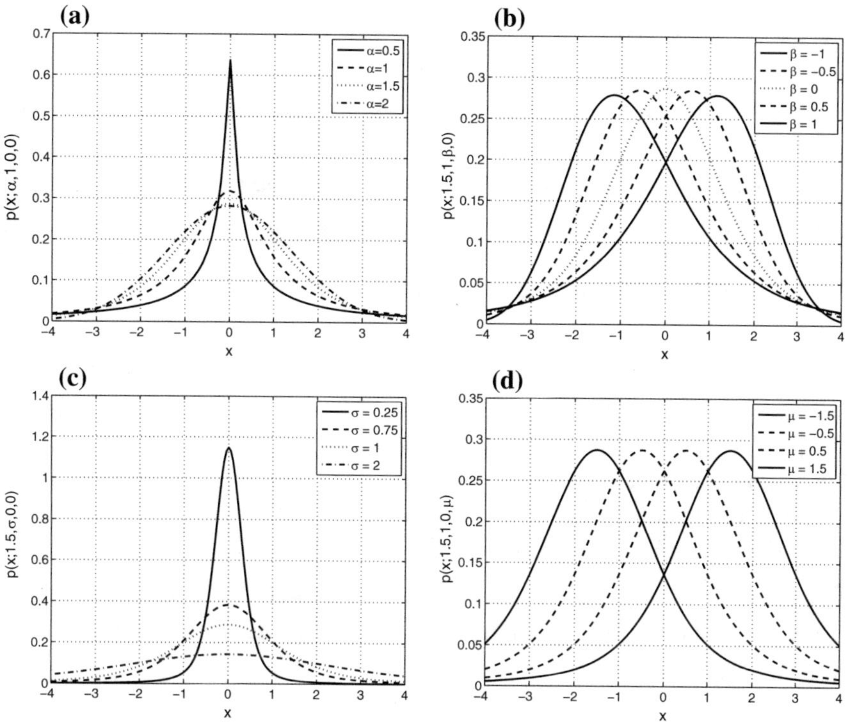

Fig. 3.1 Influence of parameters $(\alpha, \beta, \sigma, \mu)$ on the form of the probability density function

The proof can be found in many publications (e.g., Gnedenko and Kolgomorov 1954, Chap. 34). The property descibed in Theorem 3.3 is treated in many publications as a basic definition of stable distributions equivalent to Definition 3.1. The obtained family of distributions $S_\alpha(\sigma, \beta, \mu)$ are called α-stable distributions or Lévy stable distributions. A distribution from this family is defined by four parameters $(\alpha, \beta, \sigma, \mu)$; Fig. 3.1 illustrates their influence on the probability density function form.

Unfortunately, only three α-stable distributions possess the analytical form of the probability density function. These include

- normal (Gaussian) distribution ($X \sim S_2(\sigma, 0, \mu) = N(\mu, 2\sigma^2)$):

$$p_G(x) = \frac{1}{2\sigma\sqrt{\pi}} \exp\left(-\frac{(x-\mu)^2}{4\sigma^2}\right), \quad -\infty < x < \infty; \tag{3.6}$$

- Cauchy distribution ($X \sim S_1(\sigma, 0, \mu) = C(\mu, \sigma)$):

$$p_C(x) = \frac{1}{\pi} \frac{\sigma}{\sigma^2 + (x-\mu)^2}, \quad -\infty < x < \infty; \tag{3.7}$$

- Lévy distribution ($X \sim S_{1/2}(\sigma, 1, \mu) = Levy(\mu, \sigma)$):

$$p_L(x) = \sqrt{\frac{\sigma}{2\pi}} \frac{1}{(x-\mu)^{3/2}} \exp\left(-\frac{\sigma}{2(x-\mu)}\right), \quad \mu < x < \infty. \tag{3.8}$$

The greater part of α-stable distribution applications described in this book concerns symmetric α-stable distributions $S\alpha S(\sigma)$, for which $X \stackrel{d}{=} -X$. These distributions are obtained from $S_\alpha(\sigma, \beta, \mu)$ with $\beta = 0$ and $\mu = 0$, which means that $S\alpha S(\sigma) = S_\alpha(\sigma, 0, 0)$, and the characteristic function of these distributions reduces to the form

$$\varphi(k) = \exp\left(-\sigma^\alpha |k|^\alpha\right). \tag{3.9}$$

It is easy to prove that the symmetric α-stable random variable is strictly stable. The distribution $S\alpha S = S\alpha S(1)$ is called the standard symmetric α-stable distribution. It is worth noting that the standard distribution $S\alpha S$ for $\alpha = 2$ corresponds with the normal distribution $N(0, 2)$. Another interesting property is described by the theorem below (Samorodnitsky and Taqqu 1994, Proposition 1.3.1).

Theorem 3.4 *Let $X \sim S_{\alpha'}(\sigma, 0, 0)$ for $0 < \alpha' \leqslant 2$ and $0 < \alpha < \alpha'$. Let $A \sim S_{\alpha/\alpha'}\left(\left(\cos\left(\frac{\pi\alpha}{2\alpha'}\right)\right)^{\alpha'/\alpha}, 1, 0\right)$ be an extremely skew random variable independent of X. Then*

$$Z = A^{1/\alpha'} X \sim S_\alpha(\sigma, 0, 0). \tag{3.10}$$

One of particular remarks following from the above theorem is that, if $X \sim N(0, \sigma^2)$ and A is a positive $\frac{\alpha}{2}$-stable random variable independent of X, then

$$Z = A^{1/2} X \sim S\alpha S(\sigma). \tag{3.11}$$

3.1.2 Chosen Properties of Stable Distributions

The classical formula of the central limit theorem states that the limit of the normalized sum of independent random variables of an identical distribution (i.i.d.) and finite variance is the normal distribution. If we give up the assumption of the finiteness of the variance of i.i.d. variables, then the generalized central limit theorem is valid. It states that some stable distribution is a limit of the above-mentioned sum (Gnedenko and Kolgomorov 1954, p.162).

Theorem 3.5 *Let $X_1, X_2, X_3, \ldots$, be a sequence of i.i.d. random variables. Then there exist constants $a_n > 0$, $b_n \in \mathbb{R}$ and a non-degenerated random variable Z such that*

$$a_n \left(X_1 + X_2 + \ldots + X_n\right) - b_n \longrightarrow Z \tag{3.12}$$

if and only if $Z \sim S_\alpha(\sigma, \beta, \mu)$ with some $0 < \alpha \leq 2$.

A property of great importance is described by the following theorem (Nolan 2007, Proposition 1.17).

Theorem 3.6 *The parameters of the distribution* $S_\alpha(\sigma, \beta, \mu)$ (3.4) *possess the following properties:*

1. *Let* $X \sim S_\alpha(\sigma, \beta, \mu)$. *Then, for each pair of real numbers a and b* ($a \neq 0$), *we have*

$$aX + b \sim \begin{cases} S_\alpha(|a|\sigma, sign(a)\beta, a\mu + b) & \alpha \neq 1, \\ S_1(|a|\sigma, sign(a)\beta, a\mu + b - \beta\sigma\frac{2}{\pi}a\ln(a)) & \alpha = 1. \end{cases} \tag{3.13}$$

2. *The characteristic function, probability density function and cumulative distribution function are continuous for each* α *apart from* $\alpha = 1$, *but are continuous for each neighbourhood of* $\alpha = 1$.
3. *Let* $X_1 \sim S_\alpha(\sigma_1, \beta_1, \mu_1)$ *and* $X_2 \sim S_\alpha(\sigma_2, \beta_2, \mu_2)$ *be independent variables. Then* $X_1 + X_2 \sim S_\alpha(\sigma, \beta, \mu)$, *where*

$$\beta = \frac{\beta_1\sigma_1^\alpha + \beta_2\sigma_2^\alpha}{\sigma_1^\alpha + \sigma_2^\alpha}, \qquad \sigma^\alpha = \sigma_1^\alpha + \sigma_2^\alpha, \qquad \mu = \mu_1 + \mu_2. \tag{3.14}$$

In order to create an α-stable distribution $S_\alpha(\sigma, \beta, \mu)$, externally right-skewed distributions ($\beta = 1$) can be used. Such a construction procedure can be based on the following theorem (Samorodnitsky and Taqqu 1994, Property 1.2.13).

Theorem 3.7 *Let X be a stable random variable* $X \sim S_\alpha(\sigma, \beta, \mu)$ *for* $\alpha \in (0, 2)$. *Then there exist two independent random variables* Y_1 *and* Y_2 *of the same distribution* $S_\alpha(\sigma, 1, 0)$ *such that*

- $\alpha \neq 1$:

$$X \stackrel{d}{=} \left(\frac{1+\beta}{2}\right)^{1/\alpha} Y_1 - \left(\frac{1-\beta}{2}\right)^{1/\alpha} Y_2, \tag{3.15}$$

- $\alpha = 1$:

$$X \stackrel{d}{=} \left(\frac{1+\beta}{2}\right)Y_1 - \left(\frac{1-\beta}{2}\right)Y_2 + \sigma\left(\frac{1+\beta}{\pi}\ln\left(\frac{1+\beta}{2}\right) - \frac{1-\beta}{\pi}\ln\left(\frac{1-\beta}{2}\right)\right). \tag{3.16}$$

Dislocation of the probability mass from central parts of α-stable distributions to their tails with the decreased value of the stability index α is described by the following theorem (Samorodnitsky and Taqqu 1994, Property 1.2.15).

Theorem 3.8 *Let* $X \sim S_\alpha(\sigma, \beta, \mu)$ *for* $\alpha \in (0, 2)$. *Then*

$$\lim_{x\to\infty} P(X > x) = x^{-\alpha} C_\alpha \frac{1+\beta}{2} \sigma^\alpha, \tag{3.17}$$

where

$$C_\alpha = \left(\int_0^\infty x^{-\alpha} \sin(x) dx \right)^{-1}. \tag{3.18}$$

The implication of heavy tails of stable distributions is a lack of finite moments. Properties of probability models are usually described by two parameters: the expectation value $E(X)$ and variance $\text{Var}(X) = E(X^2) - (E(X))^2$. This representation is useless in the case of stable distributions because of the following result (Samorodnitsky and Taqqu 1994, Property 1.2.16).

Theorem 3.9 *Let $X \sim S_\alpha(\sigma, \beta, \mu)$ for $\alpha \in (0, 2)$. Then*

$$E(|X|^p) = \begin{cases} < \infty \text{ for each } 0 < p < \alpha, \\ = \infty \text{ for each } \quad p \geq \alpha. \end{cases} \tag{3.19}$$

Thus, the first moment is finite for $\alpha > 1$ and equal to the localisation parameter μ (3.4). The case of $\alpha = 2$ is excluded in Theorem 3.9. Here all moments are finite.

3.1.3 Simulation of α-Stable Random Variables

Knowing analytical forms of probability density functions of Gaussian (3.6), Cauchy (3.7) and Lévy (3.8) distributions allows us to construct simple generators of random values of these distributions. Let $U, U_1, U_2 \sim U(0, 1)$ be random values of a uniform distribution on $(0, 1)$. Then

- random values
$$\begin{aligned} X_1 &= \mu + \sigma\sqrt{-2\ln(U_1)}\cos(2\pi U_2), \\ X_2 &= \mu + \sigma\sqrt{-2\ln(U_1)}\sin(2\pi U_2) \end{aligned} \tag{3.20}$$
  are independent random values of the $N(\mu, \sigma^2)$ distribution;
- a random value of the Cauchy distribution $C(\mu, \sigma)$ can be obtained as follows:
$$X = \sigma \tan(\pi(U - 1/2)) + \mu; \tag{3.21}$$
- a random value of the Lévy distribution $Levy(\mu, \sigma)$ is represented by
$$X = \sigma \frac{1}{Z^2} + \mu, \tag{3.22}$$

where $Z \sim N(0, 1)$.

In the general case, the generation of random variables of α-stable distributions is more complicated because of the lack of the analytical form of inverse functions of the cumulative distribution function, apart from, of course, the normal, Cauchy and Lévy distributions described above. The first step in building a generator of stable random values was made by Kanter (1975), who showed an indirect method of simulation of random values of the $S_\alpha(1, 1, 0)$ distribution for $\alpha < 1$. The general scheme of stable random values was proposed by Chambers et al. (1976), while the description presented below is cited from the book of Nolan (2007, Theorem 1.19).

Theorem 3.10 *Let V and W be independent variables: $V \sim U\big(-\frac{\pi}{2}, \frac{\pi}{2}\big)$, and W be of an exponential distribution with the expectation value equal to one, $0 < \alpha \leq 2$.*

1. *The symmetric random variable*

$$Z = \begin{cases} \frac{\sin(\alpha V)}{(\cos(V))^{1/\alpha}} \left[\frac{\cos((\alpha-1)V)}{W}\right]^{(1-\alpha)/\alpha} & \alpha \neq 1, \\ \tan(V) & \alpha = 1 \end{cases} \tag{3.23}$$

   is of the $S_\alpha(1, 0, 0) = S\alpha S$ distribution.

2. *In the non-symmetric case, for each $-1 \leq \beta \leq 1$, let $B_{\alpha,\beta} =$ $\arctan(\beta \tan(\pi\alpha/2))/\alpha$, where $\alpha \neq 1$. Then*

$$Z = \begin{cases} \frac{\sin(\alpha(B_{\alpha,\beta}+V))}{(\cos(\alpha B_{\alpha,\beta})\cos(V))^{1/\alpha}} \left[\frac{\cos(\alpha B_{\alpha,\beta}+(\alpha-1)V)}{W}\right]^{(1-\alpha)/\alpha} & \alpha \neq 1, \\ \frac{2}{\pi}\left[\left(\frac{\pi}{2} + \beta V\right)\tan(V) - \beta \ln\left(\frac{\frac{\pi}{2} W \cos(V)}{\frac{\pi}{2}+\beta V}\right)\right] & \alpha = 1 \end{cases} \tag{3.24}$$

   is of the $S_\alpha(1, \beta, 0)$ distribution.

Variables V and W are easy to obtain based on independent random variables of uniform distribution $U_1, U_2 \sim U(0, 1)$ and using formulae $V = \pi\left(U_1 - \frac{1}{2}\right)$ and $W = -\ln(U_2)$. Having simulation formulae for a standard stable random variable (Theorem 3.10), a stable random variable $X \sim S_\alpha(\sigma, \beta, \mu)$ for each set of α, β, σ i μ can be obtained as follows:

$$X = \begin{cases} \sigma Z + \mu & \alpha \neq 1, \\ \sigma Z + \frac{2}{\pi}\beta\sigma\ln(\sigma) + \mu & \alpha = 1, \end{cases} \tag{3.25}$$

where $Z \sim S_\alpha(1, \beta, 0)$.

3.1.4 Stable Random Variable and the Evolutionary Algorithm

What influence on a stochastic search process based on stable distributions can the lack of possibility of calculating their basic parameters (Theorem 3.9), like the variance (for $\alpha < 2$) and the expectation value (for $\alpha \leq 1$), have?

In order to answer this question, let us consider the simplest version of evolutionary strategies (1 + 1)ES (Rechenberg 1965) described in Sect. 2.2.4. Let the one-dimensional real space be a search space. In the strategy (1 + 1)ES, one descendant is created by adding a random disturbance of the normal distribution to the parent. The best solution of the pair parent–descendant is chosen as a base solution for the next random step. Let $(1+1)\mathrm{ES}_\alpha$ be an extension of the strategy considered, where the normal distribution is replaced by some symmetric α-stable distribution $S\alpha S(\sigma)$, i.e.,

$$x_N = x_{N-1} + \sigma Z_N, \tag{3.26}$$

where $Z \sim S\alpha S$.

Let us assume that, starting from some initial point $x_0 \in \mathbb{R}$, the next m mutations are finished with success, which means that the fitness of each descendant has been better than that of its parent. Then the location of the m-th individual is described by the following expression:

$$x_m = x_0 + \sigma \sum_{i=1}^{m} Z_i, \tag{3.27}$$

where $(Z_i \sim S\alpha S | i = 1, 2, \ldots, m)$ is a series of random variables realizations obtained during the searching process. According to Theorem 3.2, the random variable $\sum_{i=1}^{m} Z_i$ has a stable distribution $S\alpha S(m^{\frac{1}{\alpha}})$, thus the expectation value of the base point dislocation after m iterations $\Delta x = |x_m - x_0|$ is equal to

$$E(\Delta x) = \sigma m^{\frac{1}{\alpha}} E(|Z|). \tag{3.28}$$

In the case of $S\alpha S$ distributions with the stability index $\alpha > 1$, for which $E(|Z|) < +\infty$, the exploration range, in line with expectations, increases with the number of successful mutations m. However, in the case of $\alpha \leq 1$, the expectation value $E(|Z|) = +\infty$ and the possibility of macro-mutations, i.e., very long dislocations in the search space, strongly increase. On the one hand, this property can be much desired—the high probability of micro-mutations allows us to search a much bigger area of a searching space. On the other hand, it may be feared that the searching process will be too "chaotic", almost insensitive to a fitness function and practically deprived of exploitation abilities, i.e., those of localization of fitness function extremum points.

In order to make an attempt at exploitation abilities analysis, one can notice that, in most evolutionary algorithms, the population of alternative solutions is obtained by more than one mutation of one base solution. We can ask how many random samples of a stable distribution with a given stable index α are needed in order to have a chance to locate at most one descendant near to a base solution. Let us consider λ random variables of the $|S\alpha S(\sigma)|$ distribution which are ordered after each realization. In this way, we realize some ordered statistics $\{X_{i:\lambda} \mid i = 1, 2, \ldots, \lambda\}$, where $X_{i:\lambda} \leq X_{j:\lambda}$ for $i \leq j$. The following theorem is fulfilled (Obuchowicz and Prętki 2005).

Theorem 3.11 *Let $X \sim |S\alpha S(\sigma)|$. Then the kth moment of a random variable $X_{i:\lambda}$ exists if and only if the following inequality is fulfilled:*

$$k - \alpha(\lambda - i + 1) < 0. \tag{3.29}$$

Proof (*Prętki* 2008, Appendix A.2) Let us consider a series of random variables $(X_i \sim |S_\alpha S| \mid i = 1, \ldots, \lambda)$ and its ordered version $X_{1:\lambda} < X_{2:\lambda} < \ldots < X_{\lambda:\lambda}$. Then the random variable $X_{1:\lambda}$ can be identified with a minimal distribution of a sample of λ elements, i.e., $X_{1:\lambda} = \min\{(X_i|i = 1, \ldots, \lambda)\}$, whereas $X_{\lambda:\lambda} = \max\{(X_i|i = 1, \ldots, \lambda)\}$ can be indentified with a maximal distribution. Then the k-th moment of the random variable $X_{i:\lambda}$ can be calculated using the expression (Shao 1999)

$$E(X_{i:\lambda}^k) = \int_0^\infty x^k \, p_{\alpha,i:\lambda}(x)dx, \tag{3.30}$$

where $p_{\alpha,i:\lambda}(\cdot)$ is a probability density function of the random variable $X_{i:\lambda}$. It has the following form:

$$p_{\alpha,i:\lambda}(x) = \frac{\lambda!}{(i-1)!(\lambda-i)!}\Big[F_\alpha(x)\Big]^{i-1}\Big[1 - F_\alpha(x)\Big]^{\lambda-i} p_\alpha(x), \tag{3.31}$$

where $p_\alpha(x)$ is a probability density function of a random variable of the $|S_\alpha S|$ distribution, and $F_\alpha(x)$ is its cumulative distribution function. Both functions fulfill the following equations:

$$p_\alpha(x) = 2p_{\alpha,1}(x)I_{0,\infty}, \tag{3.32}$$
$$F_\alpha(x) = 2F_{\alpha,1}(x)I_{0,\infty}, \tag{3.33}$$

where $p_{\alpha,1}$ and $F_{\alpha,1}$ are respectively the probability density function and the cumulative distribution function of the symmetric stable random variable $S\alpha S$, while I_A is the indicator function of the set A. Based on the theory of slowly changing functions, the following expression is fulfilled:

$$\int_0^\infty g(x)dx < \infty \Leftrightarrow \forall x > 0 : \lim_{h\to\infty} \frac{g(hx)}{g(h)} = x^a,$$

where $a < -1$. Let us consider the following limit:

$$\begin{aligned}\lim_{h\to\infty} \frac{(x\,h)^k p_{i:\lambda}(x\,h)}{h^k\, p_{i:\lambda}(x)} &= x^k \lim_{h\to\infty} \frac{p_{i:\lambda}(x\,h)}{p_{i:\lambda}(x)} \\ &= x^k \lim_{h\to\infty} \left[\frac{F_\alpha(x\,h)}{F_\alpha(h)}\right]^{i-1} \left[\frac{1 - F_\alpha(x\,h)}{1 - F_\alpha(h)}\right]^{\lambda-i} \frac{p_\alpha(x\,h)}{p_\alpha(h)}.\end{aligned} \tag{3.34}$$

One can notice that

$$\lim_{h\to\infty} \frac{F_\alpha(x\,h)}{F_\alpha(h)} = 1. \tag{3.35}$$

Applying L'Hopital's rule, we have

$$\lim_{h\to\infty} \frac{1 - F_\alpha(x\,h)}{1 - F_\alpha(h)} = \lim_{h\to\infty} \frac{x\,p_\alpha(x\,h)}{p_\alpha(h)}. \tag{3.36}$$

Taking into account Theorem 3.8, we can obtain

$$\lim_{x\to\infty} p_\alpha(x) \sim \alpha C_\alpha x^{-(1+\alpha)},$$

where C_α is a constant defined in Theorem 3.8;

$$\lim_{h\to\infty} \frac{p_\alpha(x\,h)}{p_\alpha(h)} = x^{-\alpha-1}.$$

Finally, we can obtain the following:

$$\lim_{h\to\infty} \frac{(x\,h)^k\,p_{i:\lambda}(x\,h)}{h^k\,p_{i:\lambda}(h)} = x^{k-\alpha(\lambda-i+1)-1}, \tag{3.37}$$

which completes the proof. □

The conclusion of Theorem 3.11 is the following: the expectation value of the random variable $X_{1:\lambda}$ of the ordered statistics exists if $\lambda > \frac{1}{\alpha}$. This means that the local convergence of evolutionary algorithms can be effective when the number of descendants of the best-fitted parent is higher than $\frac{1}{\alpha}$. The above theorem gives the necessary condition for moments existence (especially the expectation value) of the random variable $X_{1:\lambda}$ of the ordered statistics, but their value has to be numerically calculated. Figure 3.2 presents expectation values of $X_{1:\lambda}$.

It is easy to notice (Fig. 3.2) that in the case of $\lambda < 4$ the expectation value of $X_{1:\lambda}$ increases with the stability index α decreasing. This means that an evolutionary algorithm with a low value of the stability index should have a worse ability of local optimum localization. But, in the case of $\lambda \geq 4$, we have a completely different situation. In this case distributions with heavy tails have lower expectation values of $X_{1:\lambda}$. We can conclude that evolutionary algorithms with lower stability indices can exploit the neighborhood of a local optimum with higher precision if the mean number of reproductions of a given parent is higher than 4. Simultaneously, we must remember that heavy tails generate macro-mutations more often. Thus, we hope that the class of stable distributions can be a solution to the problem of the exploitation–exploration balance of stochastic optimum search—two abilities that usually exclude each other.

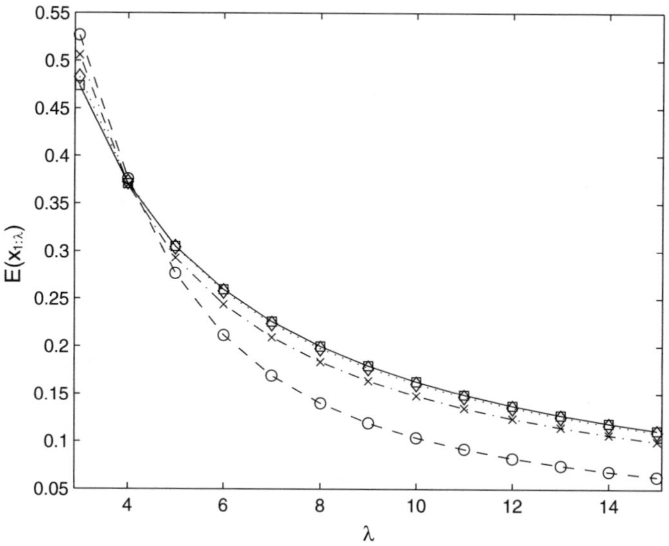

Fig. 3.2 Expectation value of the random variable $X_{1:\lambda} \sim |S\alpha S|$ versus λ ($\alpha = 2$: squares, $\alpha = 1, 5$: diamonds, $\alpha = 1$: crosses, $\alpha = 0, 5$: circles)

3.2 Stable Random Vector

3.2.1 Definition of the Stable Random Vector

The stability definition for $\mathbb{R}^n$ is similar to the respective one for $\mathbb{R}$ (Definition 3.1).

Definition 3.2 A random vector $\boldsymbol{X}$ is called a stable random vector in $\mathbb{R}^n$ (has a stable distribution in $\mathbb{R}^n$) if

$$\forall a, b > 0 \ \ \exists c > 0 \ \ \exists \boldsymbol{d} \in \mathbb{R}^n : \quad a\boldsymbol{X}^{(1)} + b\boldsymbol{X}^{(2)} \stackrel{d}{=} c\boldsymbol{X} + \boldsymbol{d}, \tag{3.38}$$

where $\boldsymbol{X}^{(1)}$, $\boldsymbol{X}^{(2)}$ are independent copies of $\boldsymbol{X}$.

If $\boldsymbol{d} = 0$ for all $a > 0$ and $b > 0$ in Definition 3.2, then we say that the random vector $\boldsymbol{X}$ is strictly stable. A random vector $\boldsymbol{X}$ has a symmetric stable distribution if it has a stable distribution and the following relation is fulfilled:

$$P\{\boldsymbol{X} \in A\} = P\{-\boldsymbol{X} \in A\} \tag{3.39}$$

for any Borel set $A \subset \mathbb{R}^n$. Like in the one-dimensional case, the symmetric stable vector is strictly stable and α-stable (Samorodnitsky and Taqqu 1994, Theorem 2.1.2).

Theorem 3.12 *Let $\boldsymbol{X}$ be a stable random vector in $\mathbb{R}^n$. Then there exists $\alpha \in (0, 2]$ such that the number c in Definition 3.2 fulfills the following equation:*

$$c = (a^\alpha + b^\alpha)^{1/\alpha} . \tag{3.40}$$

Like in the one-dimensional case (Theorem 3.3), the α-stable random vector can be defined by its characteristic function $\varphi(\boldsymbol{k}) = \mathrm{E}[\exp(-i\, \boldsymbol{k}^T \boldsymbol{X})]$. (In this book a vector is treated as a matrix composed of one column, thus $\boldsymbol{x}^T \boldsymbol{y}$ is the scalar product of vectors $\boldsymbol{x}$ and $\boldsymbol{y}$.)

Theorem 3.13 (Samorodnitsky and Taqqu 1994, Theorem 2.3.1) *Let $0 < \alpha \leqslant 2$. Then a random vector $\boldsymbol{X}$ is an α-stable random vector in $\mathbb{R}^n$ if and only if there exists a finite measure Γ_s on the unit sphere S_n in $\mathbb{R}^n$ (i.e., $S_n = \{\boldsymbol{s} : \|\boldsymbol{s}\| = 1\}$) and a vector $\boldsymbol{\mu}_0 \in \mathbb{R}^n$ such that*

- *for $\alpha \neq 1$:*

$$\varphi(\boldsymbol{k}) = \exp\left(- \int_{S_n} |\boldsymbol{k}^T \boldsymbol{s}|^\alpha \left(1 - i\,\mathrm{sgn}(\boldsymbol{k}^T \boldsymbol{s}) \tan\left(\frac{\pi\alpha}{2} \right) \right) \Gamma_s(d\boldsymbol{s}) + i\boldsymbol{k}^T \boldsymbol{\mu_0} \right), \tag{3.41}$$

- *for $\alpha = 1$:*

$$\varphi(\boldsymbol{k}) = \exp\left(- \int_{S_n} |\boldsymbol{k}^T \boldsymbol{s}| \left(1 - i\frac{2}{\pi} \mathrm{sgn}(\boldsymbol{k}^T \boldsymbol{s}) \ln|\boldsymbol{k}^T \boldsymbol{s}| \right) \Gamma_s(d\boldsymbol{s}) + i\boldsymbol{k}^T \boldsymbol{\mu_0} \right). \tag{3.42}$$

The pair $(\Gamma_s, \boldsymbol{\mu}_0)$ is a spectral representation of the vector $\boldsymbol{X}$, and Γ_s is the so-called spectral measure of the α-stable vector $\boldsymbol{X}$.

3.2.2 Selected Properties of the Stable Random Vector

One of the most important properties of stable distributions is the relation between the stability of a random vector and that of its components. The following theorem sheds some light on this problem.

Theorem 3.14 *Each linear combination*

$$Y = \boldsymbol{w}^T \boldsymbol{X} = \sum_{i=1}^{n} w_i\, X_i$$

of components of a stable random vector $\boldsymbol{X} = [X_i | i = 1, 2, \ldots, n]^T$ is a stable random variable, i.e., $Y \sim S_\alpha(\sigma_{\boldsymbol{w}}, \beta_{\boldsymbol{w}}, \mu_{\boldsymbol{w}})$, where the parameters have the following values:

$$\sigma_w = \left(\int_{S_n} |w^T s|^\alpha \Gamma_s(ds) \right)^{1/\alpha}, \tag{3.43}$$

$$\beta_w = \sigma_w^{-\alpha} \int_{S_n} \text{sign}\big(w^T s\big)|w^T s|^\alpha \Gamma_s(ds), \tag{3.44}$$

$$\mu_w = \begin{cases} w^T \mu & \alpha \neq 1, \\ w^T \mu - \frac{2}{\pi} \int_{S_n} w^T s \ln\big(|w^T s|\big)\Gamma_s(ds) & \alpha = 1. \end{cases} \tag{3.45}$$

We can conclude from the above theorem that, if a random vector is α-stable, then a random variable which is a linear combination of its components is also α-stable. Now, the following question is very interesting: Is a random vector whose components are α-stable random variables also α-stable? The answer is not unambiguous (Samorodnitsky and Taqqu 1994).

Theorem 3.15 (Samorodnitsky and Taqqu 1994, Theorem 2.1.5) *Let X be a random vector in $\mathbb{R}^n$.*

1. *If for all vectors $w \in \mathbb{R}^n$ a random variable $Y = w^T X$ has a strictly stable distribution, then X is a strictly stable random vector in $\mathbb{R}^n$.*
2. *If for all vectors $w \in \mathbb{R}^n$ a random variable $Y = w^T X$ has a symmetric stable distribution, then X is a symmetric stable vector in $\mathbb{R}^n$.*
3. *If for all vectors $w \in \mathbb{R}^n$ a random variable $Y = w^T X$ has an α-stable distribution for a stability index $\alpha \geqslant 1$, then X is an α-stable vector in $\mathbb{R}^n$.*

If we assume that the components of the random vector have a stable distribution, then their linear combination also has a stable distribution (Theorem 3.14). However, if this distribution is not strictly stable or symmetric or with the stability index $\alpha < 1$, then there is no guarantee that the random vector X is stable (Theorem 3.15).

The theorem presented below is very important for understanding the relation between a spectral measure of a random vector X and its distribution (Araujo and Gine 1980).

Theorem 3.16 *We have*

$$\lim_{r\to\infty} \frac{P\big(X \in Cone(A) \,\big|\, \|X\| > r\big)}{P\big(\|X\| > r\big)} = \frac{\Gamma_s(A)}{\Gamma_s(S_n)}, \tag{3.46}$$

where $Cone(A) = \{x \in R^n : \|x\| > 0, \frac{x}{\|x\|} \in A \subset S_n\}$, S_n i Γ_s are defined similarly as in Theorem 3.13.

The main conclusion from the above theorem is the fact that, if one knows the distribution of the spectral measure on the unit sphere, then they know the most probable directions of macro-mutations.

3.2.3 Non-isotropic Stable Random Vector

In practical applications, in mutations in evolutionary algorithms, stable random vectors based on symmetric stable distributions seem to be most attractive. The simplest method of derivation of such a stable random vector is similar to the creation of a multi-dimensional spherical Gaussian random vector $\mathbf{N}(0, \sigma^2 \boldsymbol{I}_n)$ (where $\boldsymbol{I}_n$ is a unit matrix of size $n \times n$). Thus the stable random vector of this type is composed of a set of symmetric α-stable random variables,

$$\boldsymbol{X} = [X_i \sim S\alpha S(\sigma_i)|i = 1, 2, \ldots, n]^T. \tag{3.47}$$

It may be proved that each linear combination of $\boldsymbol{X}$ components $Y = \boldsymbol{w}^T \boldsymbol{X}$ is also a random variable of a symmetric stable distribution. Thus, based on Theorem 3.15, one can say that the random vector (3.47) has an n-dimensional symmetric stable distribution. It also can be proved (Samorodnitsky and Taqqu 1994, Propositions 2.3.7–2.3.9) that the spectral representation $(\Gamma_s, \boldsymbol{\mu}_0)$ (Theorem 3.13) of the random vector $\boldsymbol{X}$ (3.47) is composed of $\boldsymbol{\mu}_0 = 0$ and a discrete spectral measure with probability mass uniformly distributed on points -1 and 1 of each axis of the Cartesian reference frame. Therefore, the distribution of the random vector $\boldsymbol{X}$ (3.47) is non-isotropic and is called the *non-isotropic symmetric* α*-stable distribution* ($\boldsymbol{X} \sim NS\alpha S(\boldsymbol{\sigma})$), or simply the *non-isotropic stable distribution*.

In order to illustrate non-isotropic stable distributions, we use the integral form of the probability density function of the random vector $\boldsymbol{X} \sim S_\alpha(\gamma, \beta, \mu)|i = 1, 2, \ldots, n]^T$ (Abdul-Hamid and Nolan 1998):

- for $\alpha \neq 1$,

$$p(\boldsymbol{x}) = \int_{S_n} g_{\alpha,n}\left(\frac{(\boldsymbol{x}-\boldsymbol{\mu})^T \boldsymbol{s}}{\sigma(\boldsymbol{s})}, \beta(\boldsymbol{s})\right)\sigma^{-n}(\boldsymbol{s})d\boldsymbol{s}, \tag{3.48}$$

where

$$g_{\alpha,n}(v, \beta) = \frac{1}{(2\pi)^n}\int_0^\infty \cos\left(v\,u - \beta\tan\left(\frac{\pi\alpha}{2}\right)u^\alpha\right)u^{n-1}\exp(-u^\alpha)\,du; \tag{3.49}$$

- for $\alpha = 1$,

$$p(\boldsymbol{x}) = \int_{S_n} g_{1,n}\left(\frac{(\boldsymbol{x}-\boldsymbol{\mu})^T \boldsymbol{s}}{\sigma(\boldsymbol{s})}, \beta(\boldsymbol{s})\right)\sigma^{-n}(\boldsymbol{s})d\boldsymbol{s}, \tag{3.50}$$

where

$$g_{1,n}(v, \beta) = \frac{1}{(2\pi)^n}\int_0^\infty \cos\left(v\,u - \frac{2}{\pi}u\,\beta\ln(u)\right)u^{n-1}\exp(-u)\,du. \tag{3.51}$$

In the case of a two-dimensional random vector $\boldsymbol{X} \sim NS\alpha S(\boldsymbol{\sigma})$, we have $\boldsymbol{x} \in \mathbb{R}^2$, $\boldsymbol{\mu} = [0, 0]^T$, $\beta(\boldsymbol{s}) = 0$ for each direction $\boldsymbol{s} \in S_2$ described on the sphere surface

S_2. The non-isotropy of the distribution of the random vector $\boldsymbol{X} \sim NS\alpha S(\boldsymbol{\sigma})$ for $\boldsymbol{\sigma} = [1, 1]^T$ is illustrated by a series of figures (Fig. 3.3). It is easy to notice that the probability mass of the density probability function focuses on the axis of the reference frame with the decreasing of the stability index. The symmetry of distributions changes from the spherical symmetry for $\alpha = 2$ to that strongly favouring the directions parallel to the axis of the reference frame for low values of α. In these cases those directions will be favoured during random sampling of the space. This effect is illustrated in Fig. 3.4, where trajectories of the two-dimensional particle disturbed by the stable random vector considered are drawn. The presented directions of macro-mutations illustrate favoured directions of random sampling for $\alpha < 2$.

Analysing the trajectories in Fig. 3.4, one can conclude that random walking of a particle disturbed by the stable vector belongs to the class of self-resembling stochastic processes, i.e., those for which the change of the process scale does not alter their statistic properties. The distribution is invariant to space scaling operations.

For making quantitative analysis of the process of probability mass concentration around axes of the reference frame (Fig. 3.3 illustrates this effect), let us consider a set of n-dimensional vectors $\{\boldsymbol{b}_k\}_{k=1}^n$ $\boldsymbol{b}_k \in \mathbb{R}^n$ defined as follows:

$$\begin{aligned}
\boldsymbol{b}_1 &= [1, 0, \ldots, 0]^T, \\
\boldsymbol{b}_2 &= [\frac{1}{\sqrt{2}}, \frac{1}{\sqrt{2}}, 0, \ldots, 0]^T, \\
\boldsymbol{b}_3 &= [\frac{1}{\sqrt{3}}, \frac{1}{\sqrt{3}}, \frac{1}{\sqrt{3}}, 0, \ldots, 0]^T, \\
\boldsymbol{b}_n &= [\frac{1}{\sqrt{n}}, \frac{1}{\sqrt{n}}, \frac{1}{\sqrt{n}}, \ldots, \frac{1}{\sqrt{n}}]^T.
\end{aligned}$$

It is worth noting that $\|\boldsymbol{b}_k\| = 1,\ k = 1, \ldots, n$.

Let the random variable

$$Y_k = \boldsymbol{b}_k^T \boldsymbol{X}_\alpha \tag{3.52}$$

be a dot product of the vector $\boldsymbol{b}_k$ and the stable vector $\boldsymbol{X}$ (3.47). Theorem 3.14 guarantees that each distribution Y_k is stable $S_\alpha(\sigma, \beta, \mu)$. If separate independent components of the vector $\boldsymbol{X}_\alpha$ have the symmetric distribution $S\alpha S(\sigma)$, then parameters of the random variable Y_k have the following properties (see Theorem 3.6): $\beta = 0$, $\mu = 0$ and

$$\sigma_{\boldsymbol{b}_k} = \sigma k^{\frac{2-\alpha}{2\alpha}}. \tag{3.53}$$

It is easy to see that the probability density function $p(y_k)$ of the dot product (3.52) in the point $y_k = 0$ is equal to the integral of the probability density function of the vector (3.47), taken over the plane perpendicular to the vector $\boldsymbol{b}_k$ and containing the origin of the reference frame:

$$p(y_k = 0) = \int_\Omega p(\boldsymbol{x}) d\boldsymbol{x}, \tag{3.54}$$

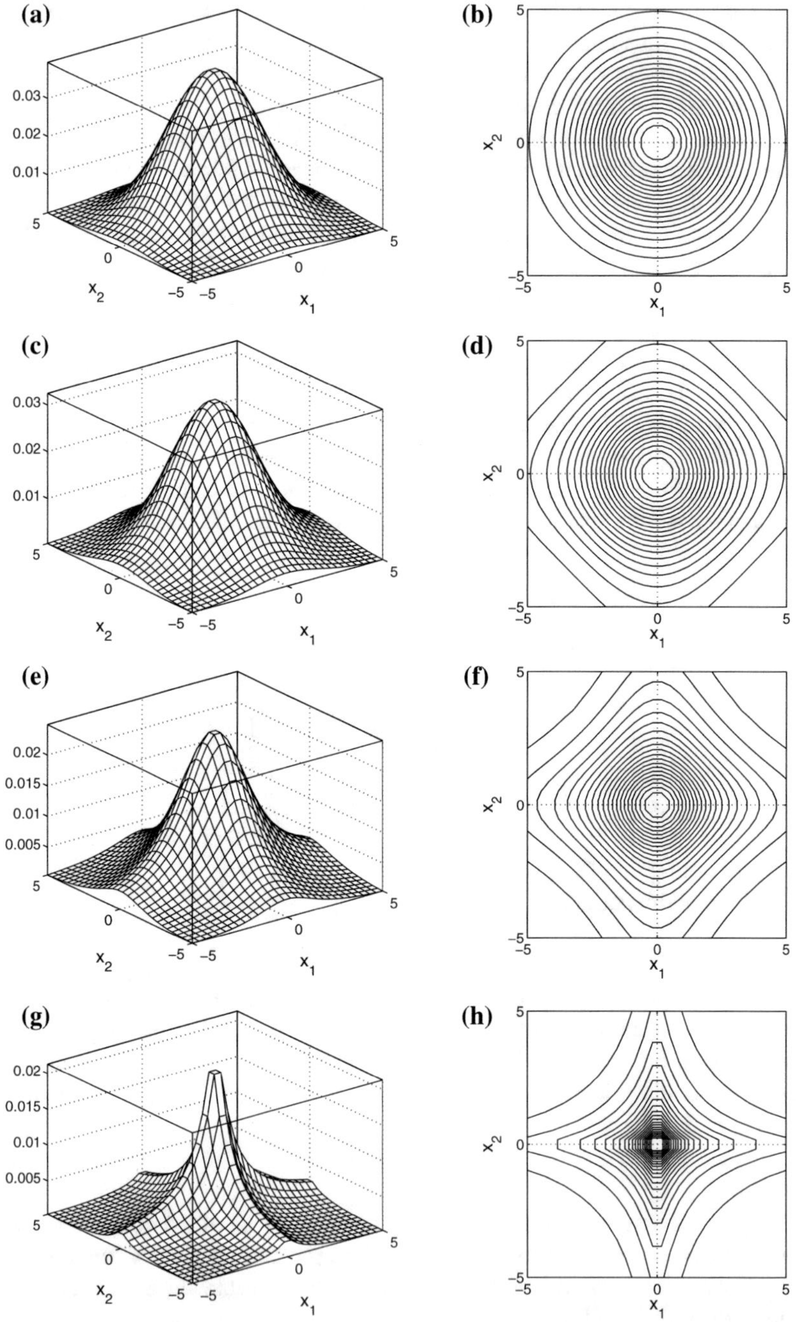

Fig. 3.3 Density probability functions of the stable random vector $\boldsymbol{X} = (X_1, X_2)$ with independent components $X_1, X_2 \sim S_\alpha S$. Stability indices: $\alpha = 2.0$ (**a**–**b**), $\alpha = 1.5$ (**c**–**d**), $\alpha = 1.0$ (**e**–**f**), $\alpha = 0.5$ (**g**–**h**)

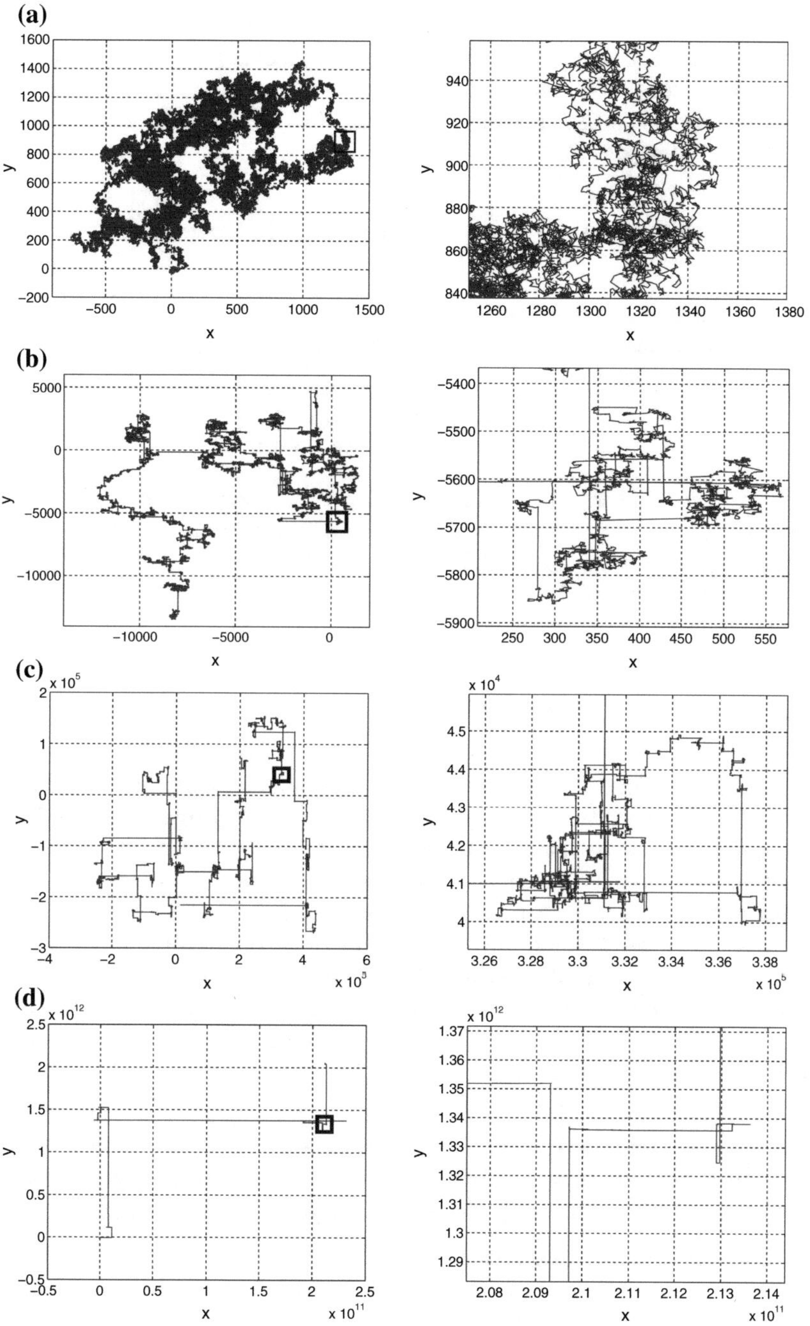

Fig. 3.4 Simulation of a random walk based on 10^6 disturbances of particles using two-dimensional non-isotropic stable distributions. Left-hand side figures show trajectories for different stable indices: $\alpha = 2$ (**a**), $\alpha = 1.5$ (**b**), $\alpha = 1$ (**c**), $\alpha = 0.5$ (**d**). Right-hand side figures show enlarged areas pointed on the right

where $\Omega = \{\boldsymbol{x} \in \mathbb{R}^n : \boldsymbol{x}^T \boldsymbol{b}_k = 0\}$. The value (3.54) does not contain information about the shape of a cross-section of the multidimensional probability density function of the vector (3.47), but about its volume in integral form. The integral (3.54) can be directly calculated knowing a characteristic function of the random variable (3.52) in the point $y_k = 0$:

$$p(y; \alpha, \sigma, 0, 0) = \frac{1}{\pi} \int_0^\infty \exp\Big(- \big(\sigma k\big)^\alpha \Big)\, dk = \frac{\Gamma(1/\alpha)}{\sigma\, \alpha\, \pi}, \tag{3.55}$$

where $\Gamma(\dot{})$ is the Euler function.

Using (3.53), the function (3.55) can be transformed to the following form:

$$p(y_k; \alpha, \sigma, 0, 0) = \frac{\Gamma(1/\alpha)}{\sigma k^{\frac{1}{\alpha} - \frac{1}{2}}\, \alpha\, \pi}. \tag{3.56}$$

Figure 3.5a–c illustrate the values of the 4-dimensional probability density function of the vector (3.47), which were calculated for the directions $\{\boldsymbol{b}_k \mid k = 1, 2, 3, 4\}$. Figure 3.5d presents the values (3.56) of the 8-dimensional stable vector (3.47) for the scale parameter $\sigma = 1$.

3.2.4 Isotropic Distributions

Isotropic distributions perform a special role in algorithms of stochastic global optimization in $\mathbb{R}^n$. In the case of searching in an unlimited real space, one can detect a slight similarity between these distributions and the uniform one within a limited set of permissible solutions. The latter distribution does not distinguish any subset of the searching space, whereas isotropic distributions do not favor any search direction. This property is the reason for the applicability of isotropic distributions at the beginning of the optimization process, when we do not have any knowledge about the problem being solved. The multidimensional isotropic distribution can be defined in, at least, a few equivalent ways. In this book we base our consideration on the theorem cited by Fang et al. (1990, Theorem 2.5).

Theorem 3.17 *Let $\boldsymbol{X} = [X_i | i = 1, 2, \ldots, n]^T$ be a random vector in $\mathbb{R}^n$, $\mathcal{O}(n)$ be a class of orthogonal matrices of the size $n \times n$, and $\boldsymbol{u}^{(n)}$ describe a random vector of a uniform distribution within the n-dimensional unit sphere. The vector $\boldsymbol{X}$ has an isotropic distribution if and only if one of the following properties is met:*

(1) $\boldsymbol{X} \overset{d}{=} \boldsymbol{O}\, \boldsymbol{X}$ for each matrix $\boldsymbol{O} \in \mathcal{O}(n)$.
(2) The characteristic function $\boldsymbol{X}$ has the form $\varphi(\boldsymbol{k}^T \boldsymbol{k})$, where $\varphi(\cdot)$ is some scalar function called a characteristic generator.
(3) $\boldsymbol{X}$ can be defined as a stochastic decomposition $\boldsymbol{X} \overset{d}{=} R\boldsymbol{u}^{(n)}$, where $R \geqslant 0$ is a real random variable (or with a distribution symmetric in respect of 0) independent of $\boldsymbol{u}^{(n)}$.

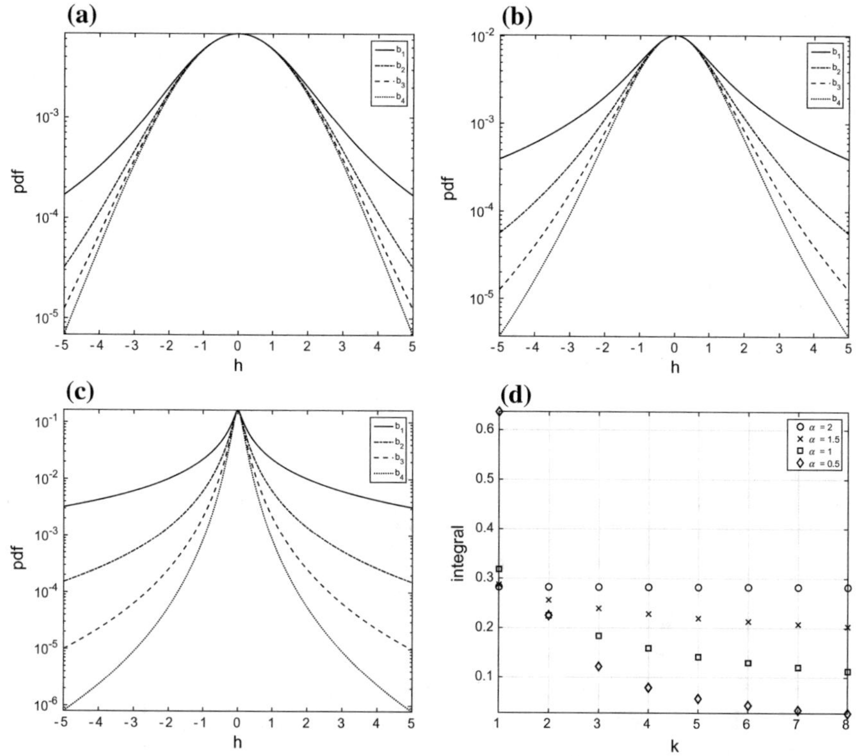

Fig. 3.5 Probability density function $p(h\boldsymbol{b}_k)$ of the stable random vector $\boldsymbol{X}_\alpha \in \mathbb{R}^4$ for $\alpha = 1.5$ (**a**), $\alpha = 1$ (**b**), $\alpha = 0.5$ (**c**). Figure (**d**) presents values $\int_{-\infty}^{\infty} p(h\boldsymbol{b}_k)dh$ for $n = 8$

(4) For each $\boldsymbol{a} \in \mathbb{R}^n$, $\boldsymbol{a}^T\boldsymbol{X} \stackrel{d}{=} \|\boldsymbol{a}\|X_1$.

It is worth noting that a generalization of a one-dimensional distribution to its multidimensional isotropic version is not trivial. Theorem 3.17 does not give us any guidance as to how to do it.

3.2.5 Isotropic Distribution Based on an α-Stable Generator

One of the possible ways to generate an isotropic random vector is direct application of the third part of Theorem 3.17. A random vector $\boldsymbol{X}$ can be defined as a decomposition

$$\boldsymbol{X} = R\boldsymbol{u}^{(n)}, \tag{3.57}$$

where R and $\boldsymbol{u}^{(n)}$ are called respectively the *generating variate* and the *uniform base of the spherical distribution*. If the generating variate is of the symmetric stable distribution $R \sim S\alpha S(\sigma)$, then we can say that the random vector $\boldsymbol{X} \sim S\alpha SU(\sigma)$ is of an *isotropic distribution based on an α-stable generator* (Obuchowicz and Prętki 2005). Probably the simplest way to generate a random vector of the uniform distribution within the surface of the unit sphere $\mathbf{u}^{(n)}$ is described by the following equations:

$$\boldsymbol{u}^{(n)} = \frac{\boldsymbol{Y}}{\|\boldsymbol{Y}\|}, \tag{3.58}$$

where $\boldsymbol{Y} \sim \mathbf{N}_n(\mathbf{0}, \boldsymbol{I}_n)$. Among many interesting properties of the vector $\boldsymbol{u}^{(n)}$, one seems to be especially appealing. The random variable $\|\boldsymbol{u}^{(n)}\|$ is degenerated, i.e., its probability mass is all accumulated in one point of the real line equal to 1. Thus, it is easy to prove that a random variable

$$Z = \|\boldsymbol{X}\| = \|R\boldsymbol{u}^{(n)}\| = |R| \tag{3.59}$$

only depends on the generating variate $|R|$ and not on the space dimension n, and the maximum of its probability density function is located in 0 (in the case of $R \sim S\alpha S(\sigma)$ for $\alpha = 2$, we obtain the half-normal distribution).

The essential property of the random vector $\boldsymbol{X}$ (3.57) is the fact that, in the general case, its multidimensional distribution is not stable.

The discussion about the probability density function of the random vector (3.57) can be started from Theorem 3.17. Taking into account the part (1) of this theorem, the probablity density function should be a function of the dot product $\boldsymbol{x}^T\boldsymbol{x}$, i.e., it should have the form $G(\boldsymbol{x}^T\boldsymbol{x})$, where the function $G(\cdot)$ is called the *probability density generator*. Formally, the probability density generator is defined as follows (Wang et al. 1997).

Definition 3.3 The scalar function $G : \mathbb{R} \to \mathbb{R}$ is called the probability density generator if it allows creating a multidimensional probability density function $p(\cdot)$ of an isotropic distribution in the following way:

$$p(\boldsymbol{x}) = c\,G(\boldsymbol{x}^T\boldsymbol{x}), \tag{3.60}$$

where c is a normalizing constant, and the following condition is met:

$$\int_0^\infty y^{n/2-1}G(y)dy < \infty. \tag{3.61}$$

The theorem described below is very useful (Fang et al. 1990, Theorem 2.9).

Theorem 3.18 *The random vector $\boldsymbol{X} = R\boldsymbol{u}^{(n)}$ has a probability density generator $G(\cdot)$ if and only if the probability density function $p(r)$ of the random variable R and the probability density generator $G(\cdot)$ fulfil the following relation:*

$$p(r) = \frac{2\pi^{n/2}}{\Gamma(n/2)}r^{n-1}G(r^2). \tag{3.62}$$

It can be concluded from the above theorem that the probability density function of the random vector obtained from the decomposition (3.57) has the form (Obuchowicz and Prętki 2005)

$$p(\boldsymbol{x}|\alpha,\sigma,n,\mu) = \frac{1}{\sigma^{(n-1)}\pi^{n/2}}\frac{\Gamma(n/2)}{\|\boldsymbol{x}-\boldsymbol{\mu}\|^{n-1}}\,p_{\alpha,1}\left(\frac{\|\boldsymbol{x}-\boldsymbol{\mu}\|}{\sigma}\right), \tag{3.63}$$

where $p_{\alpha,\sigma}(\cdot)$ is the probability density function of the random variable $R \sim S\alpha S(\sigma)$.

3.2.6 Isotropic Stable Random Vector

Let us consider an extension of the relation (3.11) defined in $\mathbb{R}$ to the multidimensional case $\mathbb{R}^n$ (Samorodnitsky and Taqqu 1994, Definition 2.5.1).

Definition 3.4 The random vector

$$\boldsymbol{X} = A^{1/2}\boldsymbol{G}, \tag{3.64}$$

where $A \sim S_{\alpha/2}\left(\left(\cos\left(\frac{\pi\alpha}{4}\right)\right)^{2/\alpha}, 1, 0\right)$ and $\boldsymbol{G} \sim \mathbf{N}(\mathbf{0}, \sigma^2\boldsymbol{I}_n)$ are stochastically independent, is called a *sub-Gaussian random vector in* $\mathbb{R}^n$.

Because each component of the random vector $\boldsymbol{X}$ (3.64) has a symmetric stable distribution $X_i = A^{1/2}G_i \sim S\alpha S(\sigma)$ (3.11), based on Theorems 3.14 and 3.15, we can conclude that the random vector $\boldsymbol{X}$ has a multidimensional symmetric α-stable distribution in $\mathbb{R}^n$. Moreover, the following theorem can be proved (Samorodnitsky and Taqqu 1994, Proposition 2.5.5).

Theorem 3.19 *Let* $\boldsymbol{X}$ *have a multidimensional symmetric* α*-stable distribution in* $\mathbb{R}^n$*. Then the following properties are equivalent:*

(a) The random vector $\boldsymbol{X}$ *is the sub-Gaussian random vector* (3.64)*, and the Gaussian component* $\boldsymbol{G}$ *in* (3.64) *is composed of i.i.d. components of the distribution* $N(0,\sigma^2)$.

(b) The characteristic function of the random vector $\boldsymbol{X}$ *has the form*

$$\varphi(\boldsymbol{k}) = \exp\left(-2^{-\alpha/2}\sigma^{\alpha}\|\boldsymbol{k}\|^{\alpha}\right). \tag{3.65}$$

(c) The random vector $\boldsymbol{X}$ *has a spectral measure which is uniformly distributed on the n-dimensional unit sphere* S_n.

If only the random vector $\boldsymbol{X}$ fulfils one of conditions described in Theorem 3.19, then it has an isotropic distribution. Thus, the distribution of the random vector $\boldsymbol{X}$ (3.64) is called the *isotropic* α*-stable distribution* ($\boldsymbol{X} \sim IS\alpha S(\sigma)$).

The stochastic decomposition (3.64) of the isotropic α-stable random vector $\boldsymbol{X}$, Theorem 3.10 and any procedure of normally distributed pseudo-random variable generation (e.g., (3.20)) can be a basis for the creation of a generator of such random vectors.

In the context of mutation operations in evolutionary algorithms, the spherical symmetry of the exploration distribution $IS\alpha S(\sigma)$, unlike in the $NS\alpha S(\boldsymbol{\sigma})$ case, fairly treats each search direction. This fact is illustrated in Fig. 3.6, which shows trajectories of a random walk consistent with the distribution $IS\alpha S(\sigma)$. Analysing these random walks, similarly as in the $NS\alpha S(\boldsymbol{\sigma})$ case (see Fig. 3.4), it is worth noticing that enlarged chosen parts of the trajectory resemble the structure of the original one.

Although the probability density function of $IS\alpha S(\sigma)$ is unknown, we can explore some knowledge from the isotropic property. Based on the known form of the characteristic function $\varphi(\boldsymbol{k})$ (3.65), we can write the probability density function in the following integral form:

$$\begin{aligned} p(\boldsymbol{x}) &= (2\pi)^{-n} \int_{\mathbb{R}^n} \exp(-i\boldsymbol{x}^T\boldsymbol{k})\varphi(\boldsymbol{k})d\boldsymbol{k} \\ &= (2\pi)^{-n} \int_{\mathbb{R}^n} \cos(\boldsymbol{x}^T\boldsymbol{k})\varphi(\boldsymbol{k})d\boldsymbol{k} - i(2\pi)^{-n} \int_{\mathbb{R}^n} \sin(\boldsymbol{x}^T\boldsymbol{k})\varphi(\boldsymbol{k})d\boldsymbol{k}. \end{aligned} \tag{3.66}$$

It is easy to show, taking into account the symmetry of integration limits and integrands, that the imaginary part of (3.66) is equal to zero. The real part can be transformed to the form (Prętki 2008, Appendix A.1)

$$\begin{aligned} &\int_{\mathbb{R}^n} \cos(\boldsymbol{x}^T\boldsymbol{k})\varphi(\boldsymbol{k})d\boldsymbol{k} = \int_0^\infty \int_{S_r^n} \cos\left(\boldsymbol{x}^T\boldsymbol{s}\right)\varphi(\boldsymbol{s})d\boldsymbol{s}\,dr \\ &= \int_0^\infty \exp\left(-2^{-\alpha/2}\sigma^\alpha r^\alpha\right) \int_{S_r^n} \cos\left(\boldsymbol{x}^T\boldsymbol{s}\right)d\boldsymbol{s}\,dr \\ &= \frac{4\pi^{\frac{n-1}{2}}}{\Gamma\left(\frac{n-1}{2}\right)} \int_0^\infty \exp\left(-2^{-\alpha/2}\sigma^\alpha r^\alpha\right)r^{n-1} \int_0^1 \cos(r\|\boldsymbol{x}\|t)\left(1-t^2\right)^{\frac{n-3}{2}} dt\,dr\,. \end{aligned}$$

Thus the probability density function of the isotropic n-dimensional stable random vector can be written in the following form:

$$p(\boldsymbol{x}) = C(n) \int_0^\infty \int_0^1 r^{n-1} \exp\left(-2^{-\alpha/2}\sigma^\alpha r^\alpha\right) \cos(r\|\boldsymbol{x}\|t)\left(1-t^2\right)^{\frac{n-3}{2}} dt\,dr, \tag{3.67}$$

where

$$C(n) = \frac{2^{2-n}}{\pi^{\frac{n+1}{2}}\Gamma\left(\frac{n-1}{2}\right)}. \tag{3.68}$$

In order to illustrate the distribution of the probability mass of the discussed random vectors in a search space, Fig. 3.7 shows cross-sections $\phi(r)$ of the probability density function (3.67) for $\sigma = 1$.

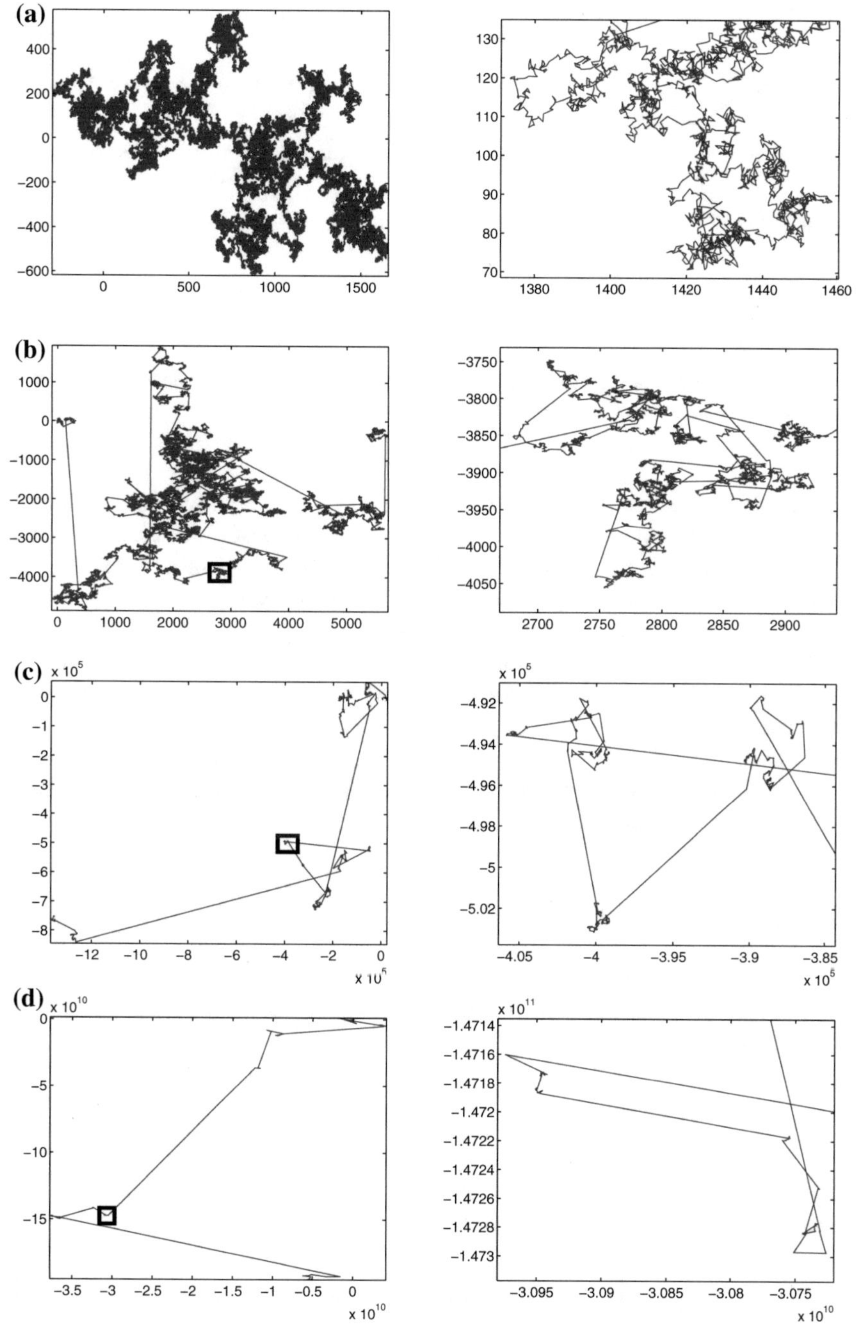

Fig. 3.6 Simulations of a random walk based on 10^6 point mutations in accordance with $IS\alpha S(\sigma)$ in a two-dimensional space. Trajectories on the left are calculated for different stable indices: $\alpha = 2$ (**a**), $\alpha = 1.5$ (**b**), $\alpha = 1$ (**c**), $\alpha = 0.5$ (**d**). Figures on the right show enlarged parts marked on the left

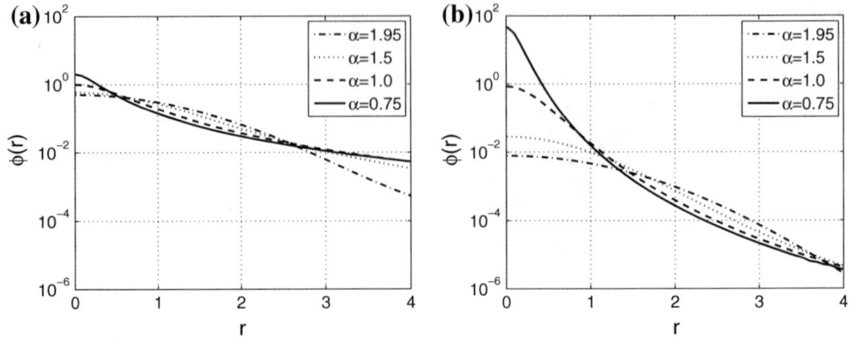

Fig. 3.7 Cross-section of the probability density function of the isotropic stable distribution for $\sigma = 1$ and two ($n = 2$ (**a**) $n = 6$ (**b**)) sample random vector dimensions

One can distinguish three areas in the figures. The probability of stable random vector location in a close neighbourhood of the beginning of the reference frame is higher for lower values of the stability index α. In the next interval of r, this relation reverses and returns to the previous one for the large values of r. This analysis suggests that, if one samples the research space in accordance with $IS\alpha S(\sigma)$, then distributions with low stability indices, on the one hand, search the neighbourhood of a base point more intensively, and on the other have a higher probability of macro-mutations.

The probability density generator (Definition 3.3) of isotropic stable distributions can be easily and directly extracted from the formulae of the probability density function of this distribution (3.67):

$$G_\alpha(k) = C(n) \int_0^\infty \int_0^1 r^{n-1} \exp\big(-2^{-\alpha/2}\sigma^\alpha r^\alpha\big)\cos(r\sqrt{kt})\big(1-t^2\big)^{\frac{n-3}{2}}\,dt\,dr, \tag{3.69}$$

with $C(n)$ defined as in (3.68).

The probability density generator can be useful for isotropic random vector determination:

$$P(\|\boldsymbol{X}_\alpha\| \leqslant R) = \int_{S_n(R)} p(\boldsymbol{x})d\boldsymbol{x} = c\int_{S_n(R)} G(\boldsymbol{x}^T\boldsymbol{x})d\boldsymbol{x}, \tag{3.70}$$

where $S_n(R)$ denotes the surface of the n-dimensional sphere of radius R. Let us transform the reference frame to the spherical form:

$$x_j = r\Big(\Pi_{k=1}^{j-1}\sin(\alpha_k)\Big)\cos(\alpha_j) \quad \text{for } 1 \le j \le n-1, \tag{3.71}$$

$$x_n = r\Big(\Pi_{k=1}^{n-2}\sin(\alpha_k)\Big)\sin(\alpha_{n-1}), \tag{3.72}$$

where $\alpha_i \in [0, \pi]$ for $i = 1, 2, \ldots, n-2$ and $\alpha_{n-1} \in [0, 2\pi)$. The Jacobian of the above reference frame transformation is equal to $r^{n-1}\Pi_{k=1}^{n-2}\sin(\alpha_k)^{n-k-1}$, and then (3.70) takes the form

$$c\int_0^R\int_0^{2\pi}\int_0^{\pi}\cdots\int_0^{\pi} r^{n-1}G(r^2)\Pi_{k=1}^{n-2}\sin(\alpha_k)^{n-k-1}drd\alpha_1 d\alpha_2\ldots d\alpha_{n-1}. \quad (3.73)$$

Taking into account the following fact:

$$\int_0^{2\pi}\int_0^{\pi}\cdots\int_0^{\pi}\Pi_{k=1}^{n-2}\sin(\alpha_k)^{n-k-1}d\alpha_1 d\alpha_2\ldots d\alpha_{n-1} = \frac{2\pi^{n/2}}{\Gamma(n/2)}, \quad (3.74)$$

we finally obtain (Prętki 2008, Appendix A.4)

$$P(\|\mathbf{Z}\| < R) = c\frac{2\pi^{n/2}}{\Gamma(n/2)}\int_0^R r^{n-1}G(r^2)dr. \quad (3.75)$$

A detailed analysis of (3.75) is described in Sect. 5.1, where the dead surrounding effect is considered (Obuchowicz 2003).

3.3 Summary

The aim of this chapter was to present the most important definitions and properties of α-stable distributions, also called in the literature Lévy stable distributions. The basic properties of stable distributions, e.g., the general central limit theorem (Theorem 3.5), the theorem on the lack of finite moments of stable distributions (Theorem 3.9) and asymptotic properties of the probability density function (Theorem 3.8), were described. Theorems which are a basis for building a generator of pseudo-random values of the α-stable distribution were presented (Theorem 3.10). One of the most important results of this chapter is the proof of the theorem on necessary conditions for the existence of finite moments of ordered stable statistics (Theorem 3.11). The discussion of this theorem allows hypothesizing that the expected effect of the application of α-stable distributions with a low stability index increases their exploration abilities (heavy tails guarantee a higher probability of macro-mutations) and their exploitation abilities—two apparently exclusive properties of stochastic optimization algorithms.

Next, multidimensional distributions were introduced and analyzed. Three types of random vectors based on α-stable distributions: the non-isotropic symmetric stable distribution $NS\alpha S(\boldsymbol{\sigma})$, the isotropic distribution with an α-stable generator $S\alpha SU(\sigma)$ and the isotropic stable distribution $IS\alpha S(\sigma)$ were presented and shortly characterized.

References

Abdul-Hamid, H., & Nolan, J. P. (1998). Multivariate stable densities as functions of one-dimentional projections. *Journal of Multivariate Analysis*, *67*(1), 80–89.

Araujo, A., & Gine, E. (1980). *The central limit theorem for real and banach valued random variables*. New York: Wiley.

Chambers, J. M., Mallows, C., & L., & Stuck, B.W., (1976). A method for simulating stable random variables. *Journal of the American Statistical Association*, *71*(354), 340–344.

Fang, K.-T., Kotz, S., & Ng, K. W. (1990). *Symmetric multivariate and related distributions*. London: Chapman and Hall.

Gnedenko, B.V., and Kolgomorov, A.N. (1954). Limit distributions for sums of independent random variables. Reading: Addison-Wesley.

Feller, W. (1971). *An introduction to the probability theory and its applications*. New York: Wiley.

Kanter, M. (1975). Stable densities under change of scale and total variation inequalities. *The Annals of Probability*, *3*, 687–707.

Nolan, J. P. (2007). *Stable distributions-models for heavy tailed data*. Boston: Birkhäuser.

Obuchowicz, A. (2003). *Evolutionary algorithms in global optimization and dynamic system diagnosis*. Zielona Góra: Lubuskie Scientific Society.

Obuchowicz, A., & Prętki, P. (2005). Isotropic symmetric α-stable mutations for evolutionary algorithms. In I. E. E. E. Congress (Ed.), *on Evolutionary Computation* (pp. 404–410). UK: Edinbourgh.

Prętki, P. (2008). *α-stable distributions in evolutionary algorithms for global parameter optimization - Ph.D thesis*. University of Zielona Góra (in Polish)

Rechenberg, I. (1965). *Cybernetic solution path of an experimental problem*. Royal Aircraft Establishment, Library Translation 1122. Farnborough: Hants.

Samorodnitsky, G., & Taqqu, M. S. (1994). *Stable non-gaussian random processes*. New York: Chapman and Hall.

Shao, J. (1999). *Mathematical Statistics*. New York: Springer.

Wang, G., Goodman, E.D., Punch, W.F. (1997). Toward the optimization of a class of black box optimization algorithms. In *9th International Conference on Tools with Artificial Intelligence* (pp. 348–360). USA: Washington.

Chapter 4
Non-isotropic Stable Mutation

The most popular mutation operation in evolutionary algorithms based on a real representation of individuals is addition to each component of the parent individual a normally distributed random value. Mutations in phenotypic evolutionary algorithms (see Chap. 2), like evolutionary programming (Table 2.2), evolutionary strategies (2.12) and evolutionary search of soft selection (2.13), are typical realizations of the above approach. Therefore, the most natural application of α-stable distributions to the mutation operation is exchanging the additive Gaussian disturbance of each component x_i of a parent vector by a random variable of the symmetric α-stable distribution $X_i \sim S\alpha S$ (one can calculate such a random variable based on Theorem 3.10), which is controlled by the scale parameter σ_i:

$$x_i' = x_i + \sigma_i X_i, \quad i = 1, 2, \ldots, n, \tag{4.1}$$

where n is a searching space dimension. In other words, the descendant vector $\boldsymbol{x}'$ is obtained by adding some random vector $\boldsymbol{X} \sim NS\alpha S(\boldsymbol{\sigma})$ (3.47) for $\boldsymbol{\sigma} = [\sigma_i \mid i = 1, 2, \ldots, n]^T$ to the current base vector $\boldsymbol{x}$:

$$\boldsymbol{x}' = \boldsymbol{x} + \boldsymbol{X}. \tag{4.2}$$

The operation (4.2) is called *non-isotropic stable mutation*.

An essential element of the exploitation ability of global optimization stochastic algorithms, like evolutionary ones, is the algorithm's ability of accurate sampling of the current base point surrounding. Stochastic analysis of the obtained distances between the base and descendant points $\|\boldsymbol{x} - \boldsymbol{x}'\|$ seems to be very important. It appears (Obuchowicz 2001) that, if the mutation (4.2) is applied, then the probability of descendant allocation near the base point rapidly decreases with the increase of the search space dimension—the so-called *dead surrounding effect* comes into being (see Sect. 4.1). The consequence of the lack of a spherical symmetry of $NS\alpha S(\boldsymbol{\sigma})$ is existence of a strong dependence between evolutionary algorithm effectiveness and reference frame selection (*symmetry effect*; Sect. 3.2.3).

A. Obuchowicz, *Stable Mutations for Evolutionary Algorithms*,
Studies in Computational Intelligence 797,
https://doi.org/10.1007/978-3-030-01548-0_4

This chapter presents a detailed discussion of the influence of both effects on an evolutionary process.

4.1 Dead Surrounding Effect

The dead surrounding effect, which seems to be unnoticed by researchers, is connected with the probability distribution of distances $\|\boldsymbol{x} - \boldsymbol{x}'\|$ between the parent element $\boldsymbol{x}$ and the descendant $\boldsymbol{x}'$, which is obtained as a result of the mutation (4.2). Figure 4.1 shows histograms of distances between the base element and its 10^6 descendants obtained using the mutation (4.2) for different searching space dimensions n and different stable indices. However, the density probability function of the non-isotropic stable random vector $\boldsymbol{X} \sim NS\alpha S(\boldsymbol{\sigma})$ possesses its maximum localized in the base point; it can be shown (Obuchowicz 2001) that the most probable distance $\|\boldsymbol{x} - \boldsymbol{x}'\| = \|\boldsymbol{X}\|$ (4.2) is equal to zero only in the one-dimension case. In the multidimensional one the most probable distance d increases with the space dimension n. The multidimensional normal distribution $\mathbf{N}(\mathbf{0}, \boldsymbol{I}_n)$ is a representative sample. In this case, the norm of the random vector $\boldsymbol{X} \sim \mathbf{N}(\mathbf{0}, \boldsymbol{I}_n)$ has the well-known distribution $\|\boldsymbol{X}\| \sim \chi_n$ (Shao 1999), for which an expectation value $\sqrt{2}\frac{\Gamma(\frac{n+1}{2})}{\Gamma(\frac{n}{2})}$ as well as the distribution mode $\sqrt{n-1}$ is easy to calculate. It is worth noting that the random vector $\boldsymbol{X} \sim \mathbf{N}(\mathbf{0}, \boldsymbol{I}_n)$ can be presented in the form (Theorem 3.17)

$$\boldsymbol{X} \stackrel{d}{=} \boldsymbol{u}^{(n)} \chi_n. \tag{4.3}$$

A consequence of the above stochastic decomposition is the fact that the degree of similarity between the parent $\boldsymbol{x}$ and its descendant $\boldsymbol{x}'$ solely and exclusively depends on the random variable $\|\boldsymbol{x} - \boldsymbol{x}'\| \sim \chi_n$. Taking into account the above mentioned properties of the distribution χ_n, it can be seen that the distance d increases with the space dimension n. This effect strongly influences evolutionary algorithm efficiency. Referring to one of the earliest adaptation heuristics, the so-called 1/5 success regula, it can be proved that the probability of an algorithm's success decreases with the increasing distance d (Beyer 2001). If the probability of success is not fitted to a satisfying level, then further progress of the evolutionary process is not possible. Thus, the $1/5$ success regula causes the exploration distribution to narrow in these cases. Hence one can suppose that the evolutionary algorithm's effectiveness for a local optimization task largely depends on frequent, effective mutations of small range disturbances. Analyzing the distribution χ_n, it can be shown that the range of mutation increases if so does the space size. This fact is the cause of deterioration of evolutionary algorithms effectiveness. This deterioration can reveal itself in two ways. If elitist selection is applied, then the number of iterations needed to satisfy solution localization rapidly increases. In the case of soft selections, the distance between the extremum point and the population in the selection–mutation equilibrium increases (Karcz-Dulęba 2004).

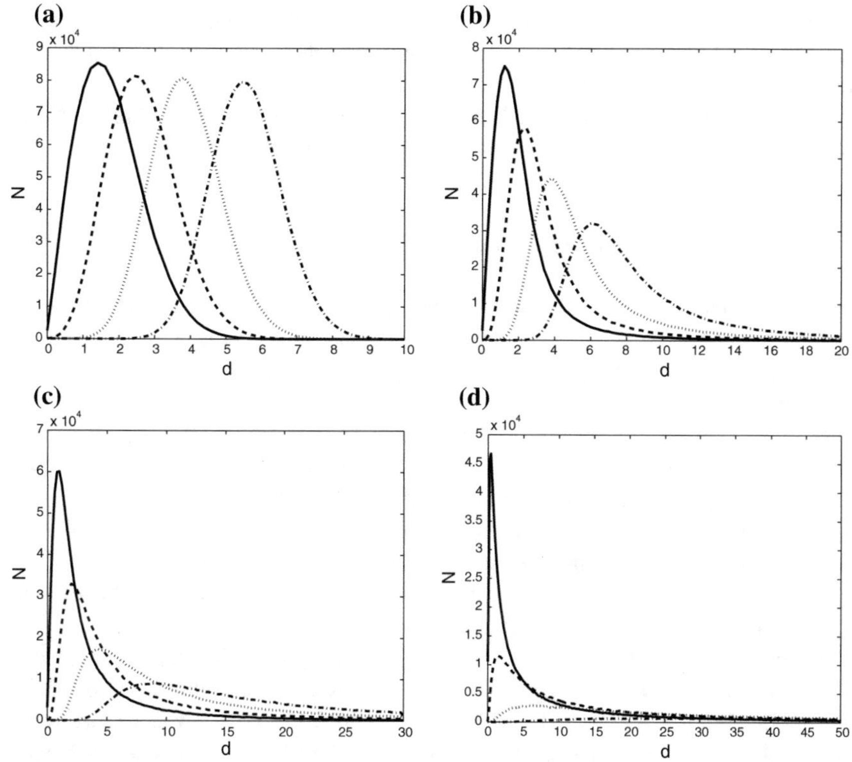

Fig. 4.1 Histograms of distances d between the base point and 10^6 points obtained using the mutation (4.2) for $\{\sigma_i = 1 \mid i = 1, 2, \ldots, n\}$ as well as $\alpha = 2$ (**a**), $\alpha = 1.5$ (**b**), $\alpha = 1$ (**c**), $\alpha = 0.5$ (**d**) ($n = 2$: solid line, $n = 4$: dashed line, $n = 8$: dotted line, and $n = 16$: dash-dot line)

4.1.1 Dead Surrounding Versus Extremum Localization

4.1.1.1 Problem Formulation

Let us consider two unimodal functions: the spherical function f_{sph} (B.1) and the generalized Rosenbrock function f_{GR} (B.2) (see Appendix B) as fitness functions.

Let $\text{ESSS}_{N;\alpha}$ and $\text{EP}_{N;\alpha}$ ($\alpha = 2, 1.5, 1, 0.5$) denote evolutionary search with soft selection (Table 2.3) and evolutionary programming (Table 2.2) algorithms in which α-stable distributions are introduced instead of the Gaussian one. In the case of the $\text{EP}_{N;\alpha}$ algorithms, mutation has the form

$$P'(t) = m_{\tau,\tau'}\big(P(t)\big) = \left\{ \boldsymbol{a}'^t_k \mid k = 1, 2, \ldots, \eta' \right\} \tag{4.4}$$

$$x'^t_{ki} = x^t_{ki} + \sigma^t_{ki} X_i \quad i = 1, 2, \ldots, n, \tag{4.5}$$

$$\sigma'^t_{ki} = \sigma^t_{ki} \exp\left(\tau' Y + \tau Y_i\right) \quad i = 1, 2, \ldots, n, \tag{4.6}$$

where $X_i, Y, Y_i \sim S\alpha S$, an index i of random variables denotes the fact that the random variable is generated independently for each component i, and a random variable Y is chosen once for all components of the vector $\boldsymbol{\sigma}_k$.

The goal of the experiment is the analysis of exploitation abilities of the algorithms considered in the sense of their convergence to the optimum point.

4.1.1.2 Experiment Description and Results

Let us consider four algorithms of the ESSS$_{N;\alpha}$ class: ESSS$_{N;2}$, ESSS$_{N;1.5}$, ESSS$_{N;1}$ and ESSS$_{N;0.5}$. Each of them is applied to the optimization process of the functions $f_{sph}(\boldsymbol{x})$ (B.1) and $f_{GR}(\boldsymbol{x})$ (B.2) for space dimensions $n = 2, 5, 10, 20$. The initial population is obtained by $\eta = 20$ mutations of a given start point $\boldsymbol{x}_0^0 = (30/\sqrt{n}, \ldots, 30/\sqrt{n})$ ($\|\boldsymbol{x}_0^0\| = 30$). The scale parameter $\sigma = 0.05$ is chosen the same for all ESSS$_{N;\alpha}$ algorithms. Each algorithm was started 500 times for each set of control parameters. The maximal number of iteration was chosen as follows: $t_{\max} = 5000$ for $f_{sph}(\boldsymbol{x})$ (B.1) and $t_{\max} = 10000$ for $f_{GR}(\boldsymbol{x})$ (B.2). Figures 4.2 and 4.3 illustrate the convergence of the best element in the current population to the optimal point for different values of the stability index α and space dimension n for $f_{sph}(\boldsymbol{x})$ and $f_{GR}(\boldsymbol{x})$, respectively.

In both cases of the fitness functions $f_{sph}(\boldsymbol{x})$ and $f_{GR}(\boldsymbol{x})$, the ESSS$_{N;\alpha}$ algorithm with low values of the stability index α converges to the extremum more quickly than others, but the extremum point is located with worse accuracy. This means that the population achieves the selection–mutation equilibrium state at a higher level of fitness function values. The accuracy of extremum localization decreases with the searching space dimension increasing. In the case of low values of the stability index α and high dimensions of the searching space, the population in the equilibrium state localizes its elements farther from the extremum than the initial point $\boldsymbol{x}_0^0$ (this effect is clearly seen in the case of the ESSS$_{N;0.5}$ algorithm).

In order to explain the above observations, let us consider the following helpful experiment, in which the spherical function $f_{sph}(\boldsymbol{x})$ (B.1) is taken into account as a fitness function. All control parameters are the same as in the previous experiment apart from the initial point, which is chosen exactly in the minimum point $\boldsymbol{x}_0^0 = (0, \ldots, 0)$. Figures 4.4 and 4.5 present the distance (in the sense of the fitness function) between the minimum and the population, which fluctuates in the selection–mutation equilibrium.

It is easy to see that the distance between the fluctuated population and the extremum point increases when so does the space dimension. This distance rapidly increases when the stability index α decreases.

The evolutionary search with soft selection algorithm ESSS$_{N;\alpha}$ does not possess the mechanism of self-adaptation of the scale parameter σ, which is the most important mutation parameter. Undoubtedly, the value of this parameter influences the range of the dead surrounding around an extremum point. It is interesting to answer the following question: Can the self-adaptation mechanism of σ included in the evolutionary algorithm overcome the *dead surrounding effect*? In order to

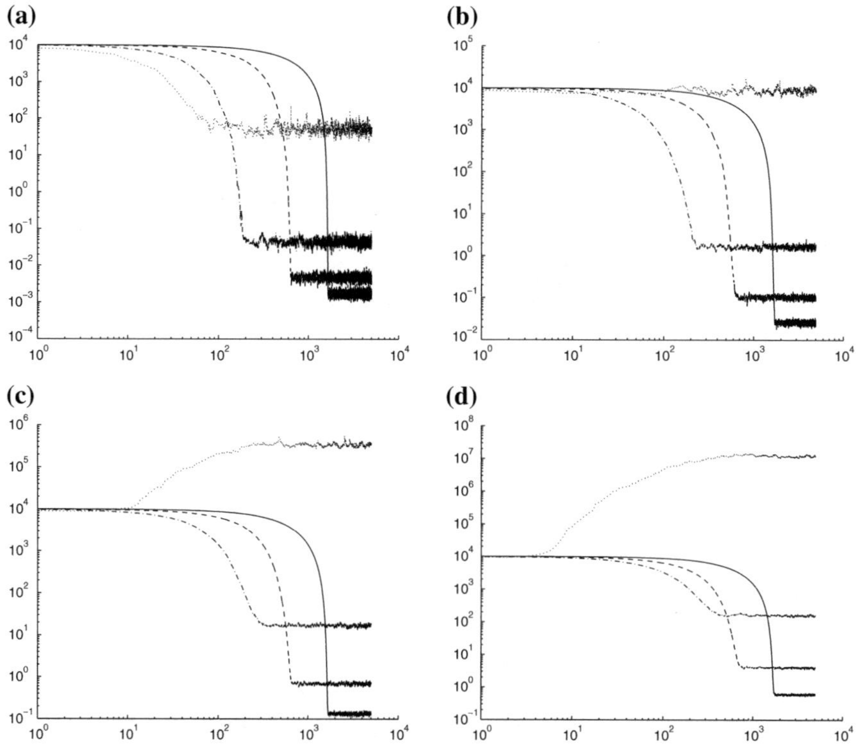

Fig. 4.2 Convergence of the $\mathrm{ESSS}_{N;\alpha}$ algorithm during the optimization process of the sphere function $f_{sph}(\boldsymbol{x})$ (B.1). The best value of $f_{sph}(\boldsymbol{x})$ in the current population versus process iterations (results average over 500 algorithm runs) for $n = 2$ (**a**), $n = 5$ (**b**), $n = 10$ (**c**), $n = 20$ (**d**) ($\alpha = 2$: solid line, $\alpha = 1.5$: dashed line, $\alpha = 1$: dash-dot line, $\alpha = 0.5$: dotted line)

answer this question, let us consider a simulation experiment. Namely, let us use four algorithms of the $\mathrm{EP}_{N;\alpha}$ class: $\mathrm{EP}_{N;2}$, $\mathrm{EP}_{N;1.5}$, $\mathrm{EP}_{N;1}$ and $\mathrm{EP}_{N;0.5}$, with the optimization process of the 2D and 20D spherical functions $f_{sph}(\boldsymbol{x})$ (B.1). The initial population of $\eta = 20$ elements was randomly selected from the limited subspace $\Omega_x = \prod_{i=1}^{n}[-10, 10]$ and $\Omega_\sigma = \prod_{i=1}^{n}[0, 0.05]$ with a uniform distribution. Other parameters are presupposed as follows: the number of sparring partners $q = 5$, the maximal number of iterations $t_{\max} = 1000$. Each algorithm was started 50 times for each set of initial parameters. Figure 4.6 presents the convergence of the best element in the current population to the extremum point for different values of α and space dimension n (results are averaged over 50 algorithm runs). The dependence between the averaged scale parameter in the current population and iterations for chosen algorithm runs is presented in Figures 4.7 and 4.8.

The convergence of the $\mathrm{EP}_{N;\alpha}$ algorithm to the extremum point (Fig. 4.6) strongly differs from the convergence of the $\mathrm{ESSS}_{N;\alpha}$ algorithm (Fig. 4.2). The $\mathrm{EP}_{N;\alpha}$ algorithm with the higher values of the stability index α more quickly converges to the extremum. Unlike in the case of the $\mathrm{ESSS}_{N;\alpha}$ algorithm, where some selection-

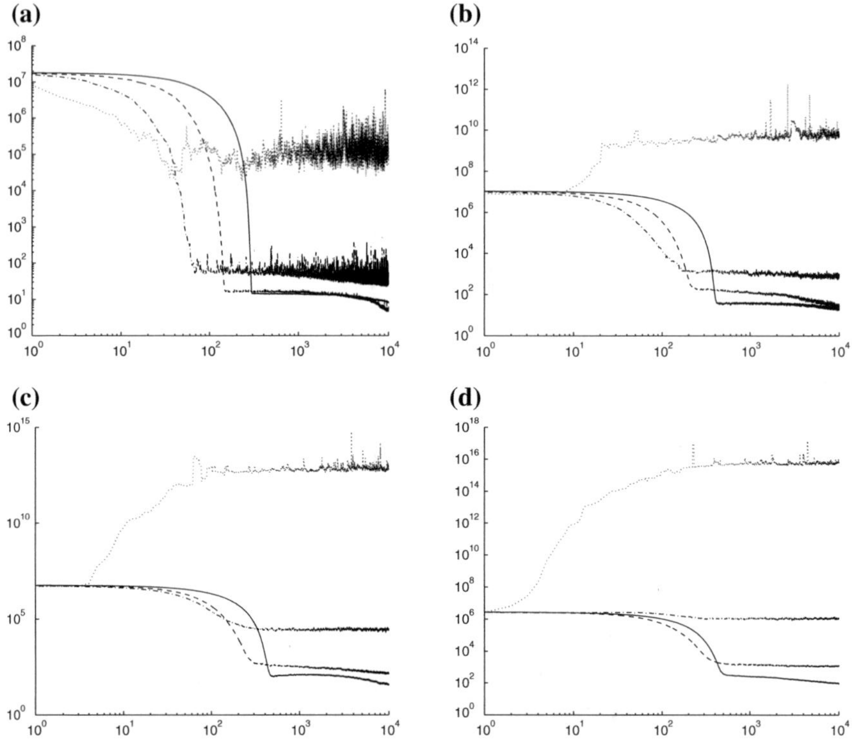

Fig. 4.3 ESSS$_{N;\alpha}$ algorithm convergence during the optimization process of the Rosenbrock function $f_{GR}(\boldsymbol{x})$ (B.2)—the best value of $f_{GR}(\boldsymbol{x})$ in the current population versus process iterations (results averaged over 500 algorithms runs) for $n = 2$ (**a**), $n = 5$ (**b**), $n = 10$ (**c**), $n = 20$ (**d**) ($\alpha = 2$: solid line, $\alpha = 1.5$: dashed line, $\alpha = 1$: dash-dot line, $\alpha = 0.5$: dotted line)

mutation equilibrium state of the population can be easily detected, the population in EP$_{N;\alpha}$ continuously and systematically converges to the local extremum. Another rule can be observed in Figs. 4.7 and 4.8. The mean value of the scale parameter (averaged over all elements in the current population) rapidly decreases—this effect is especially clearly seen for a large dimension of the searching space and possesses chaotic characteristics of changes in comparison to high values of α.

All presented results are determined by two main mechanisms. One of them is correspondence to the heavy tails of the probability density functions of symmetric α-stable distributions. They enforce population convergence to an extremum neighborhood with the decreasing of the stability index α; however, the exactness of this extremum localization is significantly worse because of the dead surrounding effect described above. This effect, connected with multi-dimensional stable distributions, does not allow locating descendants close to parents. The range of the dead surrounding increases with the space dimension increasing and the stability index decreasing.

The dead surrounding effect is recompensed by the self-adaptation mechanism of the scale parameter σ. But the reduction of the dead surrounding range is connected

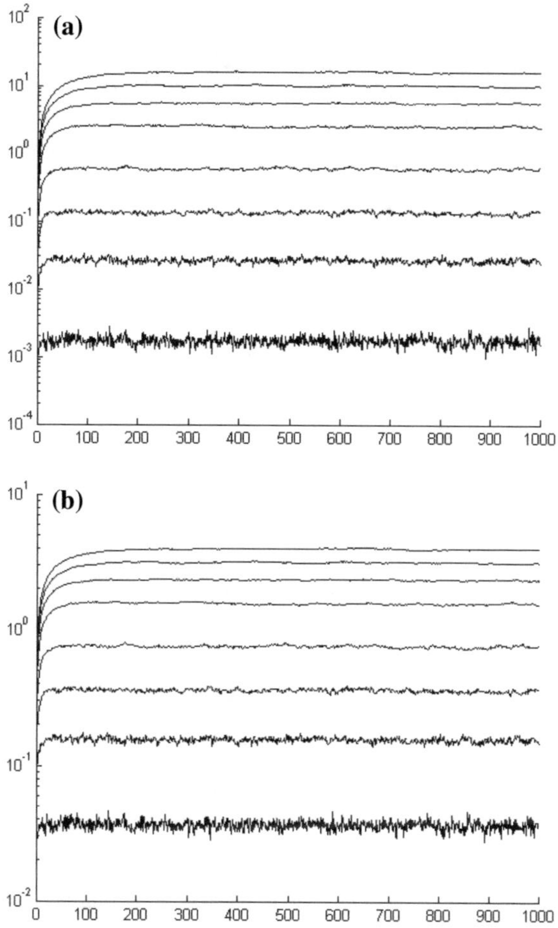

Fig. 4.4 Evolutionary process $\mathrm{ESSS}_{N;2}$ for the spherical function $f_{sph}(\boldsymbol{x})$ initiated in the extremum point—the best value of the fitness function in the current population versus iterations (**a**), as well as the distance between the best element in the current population and the optimal point in each iteration (**b**) (results are averaged over 100 algorithm runs; curves from bottom to top are obtained for space dimensions $n =$ 2, 5, 10, 20, 40, 60, 80, 100, respectively)

with σ rapidly decreasing to extremely low values, especially in the case of low values of α and high values of n. This rapid decreasing of the mutation 'range', apart from other causes, is a reason behind the fact that algorithms, despite the low values of α, converge very slowly. The exploration abilities of these algorithms also decrease.

4.1.2 Saddle Crossing

4.1.2.1 Problem Formulation

Let us consider the saddle crossing problem (Appendix A). Each algorithm considered will be started many times for a given set of control parameters. Let us consider

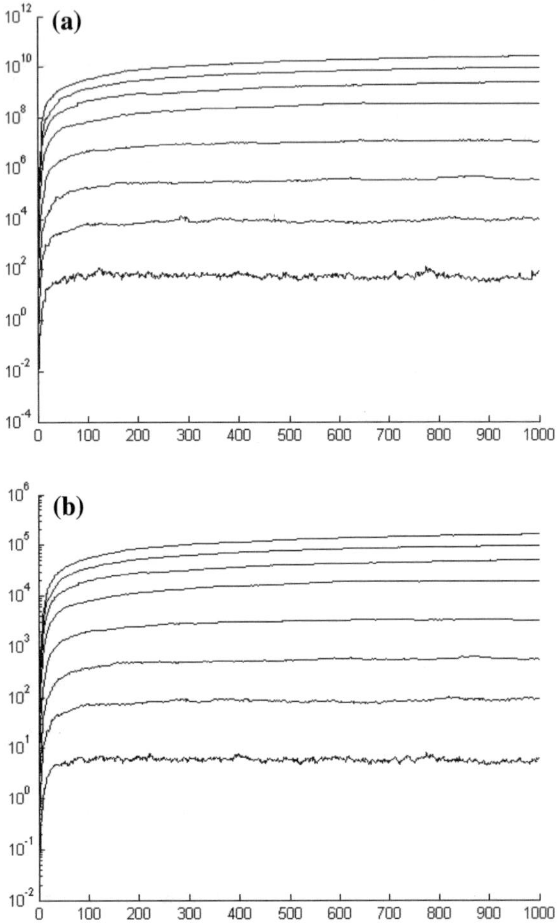

Fig. 4.5 Evolutionary process $ESSS_{N;0.5}$ for the spherical function $f_{sph}(\boldsymbol{x})$ initiated in the extremum point—the best value of the fitness function in the current population versus iterations (**a**), as well as the distance between the best element in the current population and the extremum point in each iteration (**b**) (results are averaged over 100 algorithm runs; curves from bottom to top are obtained for space dimensions $n = 2, 5, 10, 20, 40, 60, 80, 100$, respectively)

two measures of algorithm efficacy: the percentage ζ of *successful runs*, i.e., the condition (A.2) is met before the maximal number of iterations $t_{\max}$ is achieved; and the mean number of iterations $\bar{t}$, which is needed to cross a saddle by a given algorithm.

4.1.2.2 Experiment Realization and Results

Firstly, four algorithms of the $ESSS_{N;\alpha}$ class: $ESSS_{N;2}$, $ESSS_{N;1.5}$, $ESSS_{N;1}$ and $ESSS_{N;0.5}$ are analysed. The following values of control parameters are chosen: the population size $\eta = 20$, the scale parameter $\sigma = 0.05$, the maximal number of iterations $t_{\max} = 10^5$, the limit fitness (A.2) $\phi_{lim} = 0.6$. Each algorithm was started 500 times for each set of initial parameters. Results are presented in Fig. 4.9.

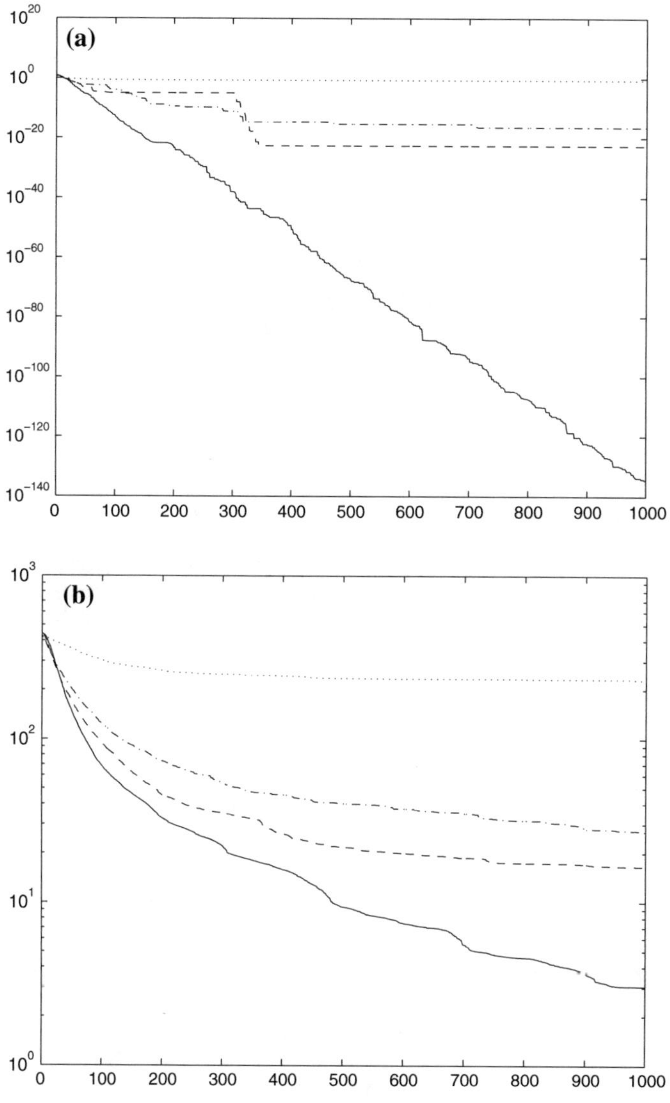

Fig. 4.6 Convergence of the $EP_{N;\alpha}$ algorithm during the optimization process of the spherical function $f_{sph}(\boldsymbol{x})$ (B.1)—the best value of $f_{sph}(\boldsymbol{x})$ in the current population versus iterations (results are averaged over 50 algorithm runs) for $n = 2$ (**a**), $n = 20$ (**b**) ($\alpha = 2$: solid line, $\alpha = 1.5$: dashed line, $\alpha = 1$: dash-dot line, $\alpha = 0.5$: dotted line)

In the case of low space dimensions n, the algorithm with the lowest stability index $\alpha = 0.5$ is most effective. The high effectiveness of $ESSS_{N;\alpha}$ with low values of α and n rapidly decreases when the search space dimension increases. In the case of high dimensions, the efficacy order of $ESSS_{N;\alpha}$ is reversed.

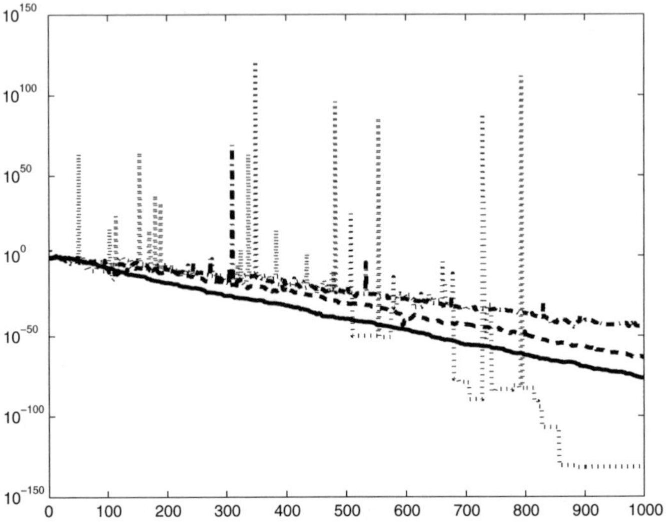

Fig. 4.7 Searching for the extremum point of the 2D spherical function $f_{sph}(\boldsymbol{x})$ (B.1) using the $EP_{N;\alpha}$ algorithm. The mean value of the scale parameter σ in the current population versus iterations ($\alpha = 2$: solid line, $\alpha = 1.5$: dashed line, $\alpha = 1$: dash-dot line, $\alpha = 0.5$: dotted line)

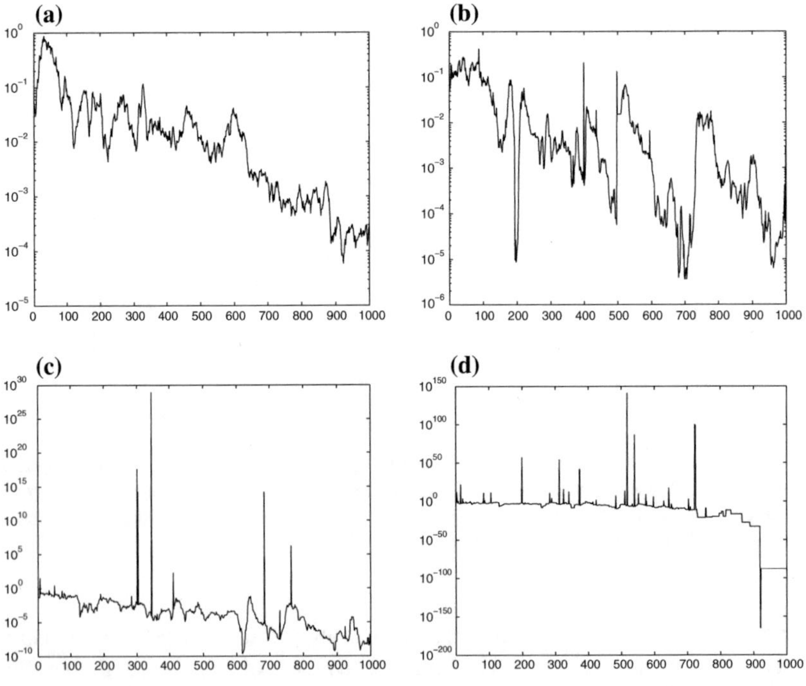

Fig. 4.8 Searching for the extremum point of the 20D spherical function $f_{sph}(\boldsymbol{x})$ (B.1) using the $EP_{N;\alpha}$ algorithm. The mean value of the scale parameter σ in the current population versus iterations ($\alpha = 2$ (a), $\alpha = 1.5$ (b), $\alpha = 1$ (c), $\alpha = 0.5$ (d))

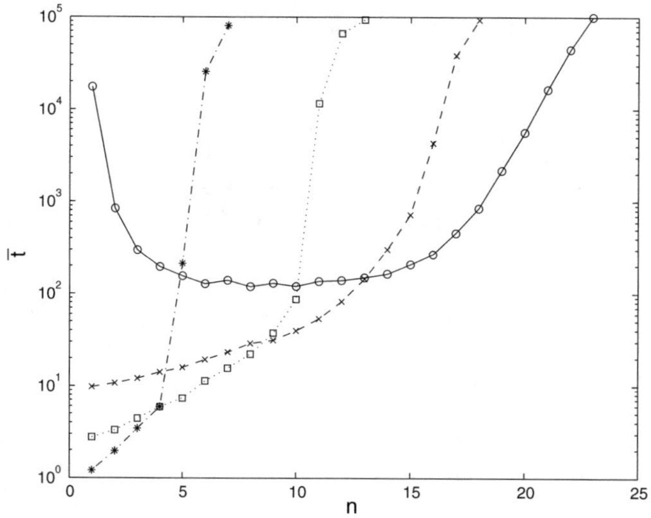

Fig. 4.9 Mean number of iterations $\bar{t}$ (taken over 500 algorithm runs) needed to cross a a saddle versus searching space dimensions n (ESSS$_{N;2}$: circles and solid line, ESSS$_{N;1.5}$: crosses and dashed line, ESSS$_{N;1}$: quarters and dotted line, ESSS$_{N;0.5}$: stars and dash-dot line)

A different relation $\bar{t}(n)$ can be observed for the ESSS$_{N;2}$ algorithm (with Gaussian mutation). The value of $\bar{t}$ is very high for $n = 1$ and decreases with n increasing until $n \approx 6$. There is some case of 'equilibrium' for higher dimensions until $n \approx 14$; in the case of $n > 14$, the relation $\bar{t}(n)$ becomes similar to relations described above for ESSS$_{N;\alpha}$ with the lowest values of α.

It is not surprising that saddle crossing effectiveness is better in the case of algorithms with the lowest values of the stability index α for low dimensions n. Both local extremum points are located on the same axis of the reference frame and the mutation probability in this direction increases when α decreases. But when the searching space dimension n increases, then the *dead surrounding effect* more significantly influences the mutation 'range'. This effect influences increase of ESSS$_{N;2}$ effectiveness with increasing n for relatively low values of n. The *dead surrounding* 'repels' mutated points from the base point and facilitates saddle crossing. This effect will be also seen in the case of ESSS$_{N;\alpha}$ for the lowest values of α if the lowest values of σ are used.

The observed rapid decrease of saddle crossing effectiveness in the case of higher space dimensions is caused by two mechanisms. The proportion between a solid angle which contains all directions of fitness function improvement (i.e., directions in which the mutations can place descendants with higher fitness), and the full n-dimensional solid angle rapidly decreases with n increasing. Then the probability of successful mutations also decreases. The second mechanism is connected with the *dead surrounding* effect. In the case of relatively low dimensions, this effect can help saddle crossing by descendant repelling from a parent, but for respectively high n the global extremum can be located inside the dead surrounding area.

In the next experiment, four algorithms of the class $EP_{N;\alpha}$: $EP_{N;2}$, $EP_{N;1.5}$, $EP_{N;1}$ and $EP_{N;0.5}$ are taken into account for the saddle crossing problem. The initial population of $\eta = 20$ elements is randomly chosen from the areas $\Omega_x = [0.8, 1.2] \times \prod_{i=2}^{n}[-0.2, 0.2]$ and $\Omega_\sigma = \prod_{i=1}^{n}[0, 0.01]$. The following algorithm control parameters are chosen: the number of sparring partners of succession $q = 5$, the maximum number of iterations $t_{\max} = 1000$ and the border fit $\phi_{lim} = 0.6$ (see (A.2)). Each algorithm was started 100 times for each set of initial parameters. Results are summarized in Fig. 4.10.

The evolutionary programming algorithm with Gaussian mutation $EP_{N;2}$ has great problems with saddle crossing in the time interval $t_{\max}$ (Fig. 4.10a). However, if the saddle is crossed, then this is done in relatively short time (Fig. 4.10b). A similar relation can be observed for $EP_{N;0.5}$ with an extremely low value of the stability index $\alpha = 0.5$.

The percentage of successful runs for $EP_{N;\alpha}$ with indices $\alpha = 1.5$ and $\alpha = 1$ is maintained at the level of $80\% < zeta \leq 100\%$. However, the effectiveness of the $EP_{N;1}$ algorithm, in the sense of $\bar{t}$, is significantly higher than in the case of $EP_{N;1.5}$.

In the case of evolutionary programming, $EP_{N;\alpha}$, the dead surrounding effect is recompensed by the self-adaptation mechanism of the scale parameter σ. Moreover, one more feature appears. When the initial values of the scale parameter are relatively low (they are randomly chosen from the area $\Omega_\sigma = \prod_{i=1}^{n}[0, 0.01]$), then the population in $EP_{N;2}$ and $EP_{N;0.5}$ either crosses the saddle in a very short time, or it cannot do it at all. If there is no mutation success at the beginning of an algorithm then the adaptation mechanism causes rapid reduction of the scale parameter value and the population gets stuck around a local extremum. The lower value of successful mutations in the case of $EP_{N;2}$ is connected with the relatively low probability of macro-mutations, while in the case of $EP_{N;0.5}$ the greater, despite heavy tails of pdf, process of scale parameter reduction is most fierce (see Figs. 4.7 and 4.8), and this fact is the cause of the population's premature convergence to a local extremum.

4.1.3 Optimization of Multimodal Functions

4.1.3.1 Problem Statement

The aim of this section is analysis of the $ESSS_{N;\alpha}$ and $EP_{N;\alpha}$ algorithms' ability of local extremum localization in chosen multimodal objective environments. Three commonly known benchmark functions are chosen as testing objective functions: Ackley's function (B.3), Rastringin's generalized function (B.4) and Griewank's generalized function (B.5). The stop criterion for all algorithms is chosen as follows:

$$\min\{\|\mathbf{x}_i^t\| \mid i = 1, 2, \ldots, \eta\} < 0.001, \tag{4.7}$$

or the maximum number of iterations $t_{\max}$ is reached.

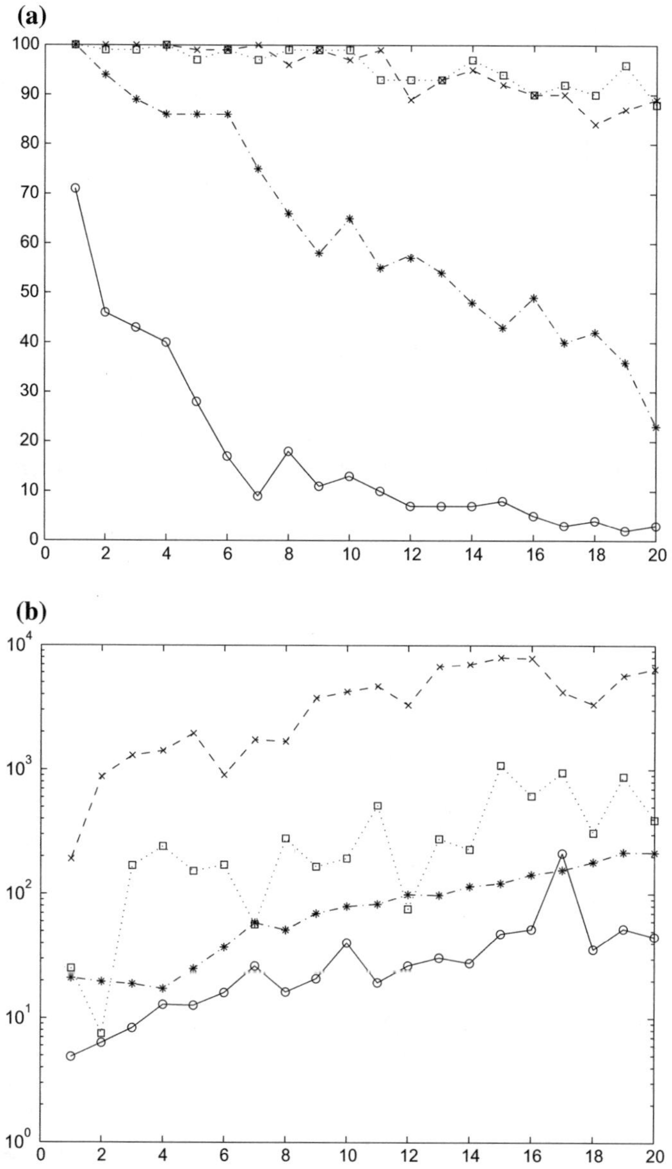

Fig. 4.10 Percentage ζ of successful algorithm runs (**a**) and the mean number of iterations $\bar{t}$ needed to cross the saddle taken over all successful algorithm runs (**b**) versus the space dimension n ($EP_{N;2}$: circles and solid line, $EP_{N;1.5}$: crosses and dashed line, $EP_{N;1}$: squares and dotted line, $EP_{N;0.5}$: stars and dash-dot line)

4.1.3.2 Experiment Realization and Results

First four algorithms of the $ESSS_{N;\alpha}$ class: $ESSS_{N;2}$, $ESSS_{N;1.5}$, $ESSS_{N;1}$ and $ESSS_{N;0.5}$ are analyzed. Each of them is applied to the optimization process of 2D, 5D, 10D and 20D versions of the $f_A(\boldsymbol{x})$ (B.3), $f_R(\boldsymbol{x})$ (B.4) and $f_G(\boldsymbol{x})$ (B.5) functions. The initial population is obtained by $\eta = 20$ mutations of the start point $\boldsymbol{x}_0^0 = (30, 30, 0, \ldots, 0)$. The scale parameter is the same for all algorithm runs: $\sigma = 0.05$. Each algorithm is started 50 times for each set of initial parameters. The maximum number of iterations is chosen as $t_{max} = 100000$. Figures 4.11, 4.12 and 4.13 illustrate the convergence of the best elements in the current population to the global extremum for different values of α and n.

Populations in the $ESSS_{N;0.5}$ algorithm cross the saddle between valleys of the 2D function faster; however, the precision of global extremum localization is the worst in comparison to the $ESSS_{N;1.5}$ and $ESSS_{N;1}$ algorithms. The effectiveness of $ESSS_{N;0.5}$ rapidly decreases with the search space dimension increasing. The

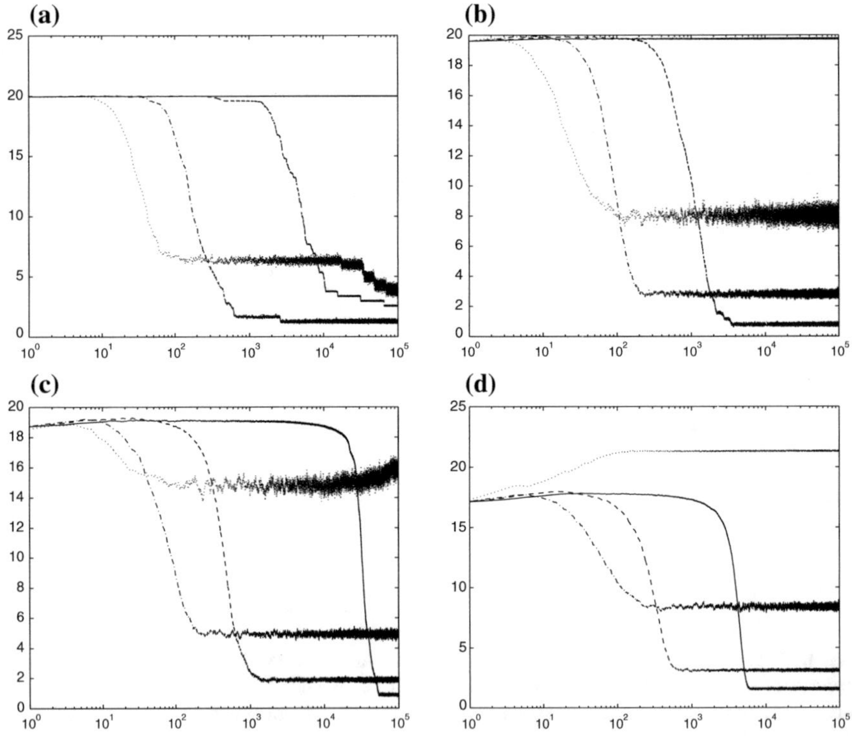

Fig. 4.11 Ackley's function $f_A(\boldsymbol{x})$ (B.3) optimized by the $ESSS_{N;\alpha}$ algorithm. Value $f_A(\boldsymbol{x})$ of the best element in the current population versus iterations (results averaged over 50 algorithm runs) for space dimensions $n = 2$ (**a**), $n = 5$ (**b**), $n = 10$ (**c**) and $n = 20$ (**d**) ($\alpha = 2$: solid line, $\alpha = 1.5$: dashed line, $\alpha = 1$: dash-dot line and $\alpha = 0.5$: dotted line)

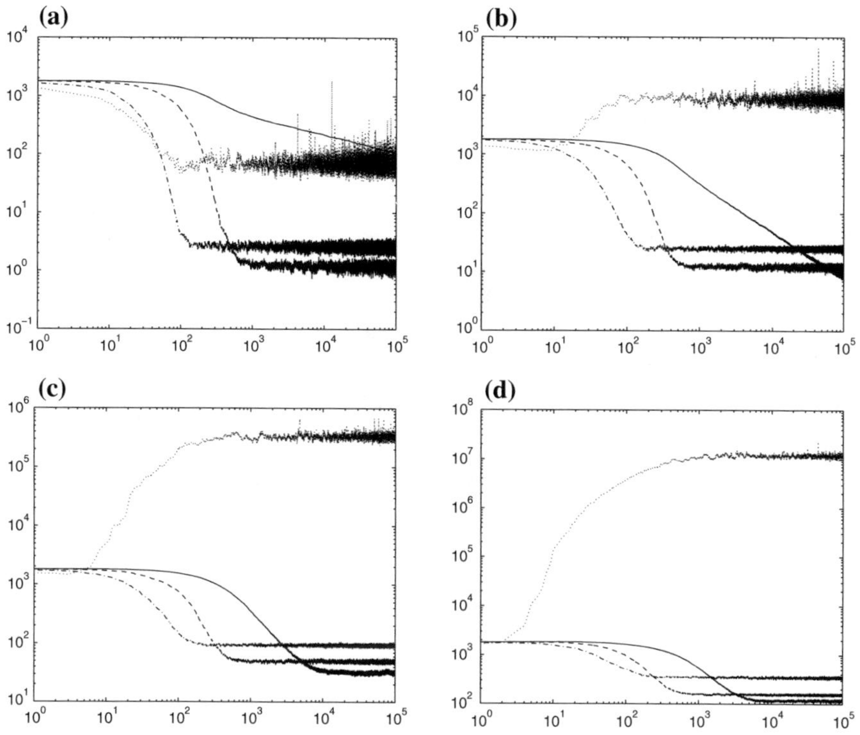

Fig. 4.12 Rastringin's function $f_R(\boldsymbol{x})$ (B.4) optimized by the $\mathrm{ESSS}_{N;\alpha}$ algorithm. Value $f_R(\boldsymbol{x})$ of the best element in the current population versus iterations (results averaged over 50 algorithm runs) for space dimensions $n = 2$ (**a**), $n = 5$ (**b**), $n = 10$ (**c**) and $n = 20$ (**d**) ($\alpha = 2$: solid line, $\alpha = 1.5$: dashed line, $\alpha = 1$: dash-dot line and $\alpha = 0.5$: dotted line)

population fluctuates around the selection-mutation equilibrium on a level worse than the beginning point $\boldsymbol{x}_0^0$.

The $\mathrm{ESSS}_{N;2}$ algorithm gives completely different results than $\mathrm{ESSS}_{N;0.5}$. In this case, the population has serious problems with escape from the local valley around the start point $\boldsymbol{x}_0^0$. The saddle crossing ability of the population in $\mathrm{ESSS}_{N;2}$ increases with the space dimension n increasing. However, the convergence to the local extremum is slower than in the case of $\mathrm{ESSS}_{N;1.5}$ and $\mathrm{ESSS}_{N;1}$, but this point is localized more correctly than in other algorithms (in the case of Ackley's and Rastrinigen's functions as well as $n \in \{5, 10, 20\}$).

The $\mathrm{ESSS}_{N;1.5}$ and $\mathrm{ESSS}_{N;1}$ algorithms combine both of the abilities described above and have a good effect in all cases; however, this effectiveness decreases with the space dimension n increasing.

In the second part of the experiment, 16 algorithms of the $\mathrm{EP}_{N;\alpha}$ class with the stability indices $\alpha = 2.0, 1.9, 1.8, \ldots, 0.6, 0.5$ are used. The initial population of $\eta = 20$ individuals was randomly chosen from the subspace $\Omega_x = \prod_{i=1}^{2}[29, 31] \times \prod_{i=3}^{n}[-0.05, 0.05]$ and $\Omega_\sigma = \prod_{i=1}^{n}[0, 0.05]$. The number of sparing partners of the selection operator is $q = 5$. The maximum number of iterations is chosen as

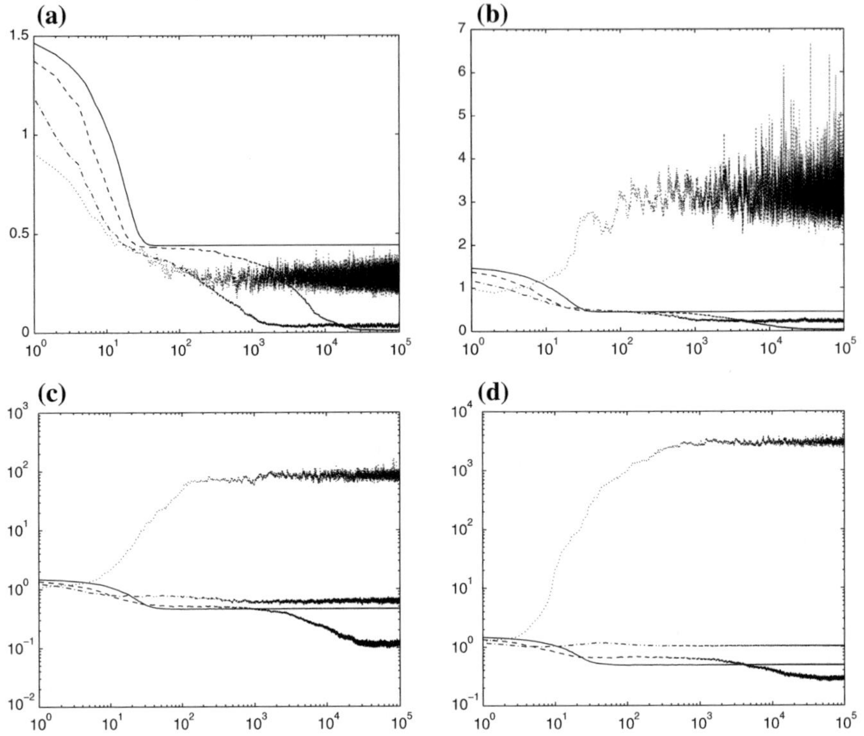

Fig. 4.13 Griewank's function $f_G(\boldsymbol{x})$ (B.5) optimized by the $\text{ESSS}_{N;\alpha}$ algorithm. Value $f_G(\boldsymbol{x})$ of the best element in the current population versus iterations (results averaged over 50 algorithm runs) for space dimensions $n = 2$ (**a**), $n = 5$ (**b**), $n = 10$ (**c**) and $n = 20$ (**d**) ($\alpha = 2$: solid line, $\alpha = 1.5$: dashed line, $\alpha = 1$: dash-dot line and $\alpha = 0.5$: dotted line)

$t_{\max} = 100000$. Each algorithm was started 50 times for each set of initial parameters. Although all algorithms were used for optimization of the functions $f_A(\boldsymbol{x})$ (B.3), $f_R(\boldsymbol{x})$ (B.4) and $f_G(\boldsymbol{x})$ (B.5) in 2D, 5D, 10D and 20D spaces, only the results for Rastringin's function $f_R(\boldsymbol{x})$ are presented in Figs. 4.14, 4.15, 4.16 and 4.17, because results for all the functions considered possess similar properties.

The percentage ζ of successful algorithm runs decreases with the space dimension increasing, and this tendency is independent of the stability index and the objective function.

The worst results were obtained for classical evolutionary programming $\text{EP}_{N;2}$. The $\text{EP}_{N;\alpha}$ algorithm for $1 \leq \alpha \leq 1.5$ gives the best results for all searching space dimensions considered. In the case of $\alpha < 1$, ζ decreases with α. Application of $\text{EP}_{N;\alpha}$ for $\alpha < 0.5$ has stopped without success for all searching space dimensions.

The above observations suggest that the *dead surrounding* effect is one of the most important mechanisms influencing evolutionary algorithms' effectiveness in multimodal function global optimization problems. The $\text{ESSS}_{N;0.5}$ algorithm is not effective because of the existence of a wide dead area. The same effect enforces

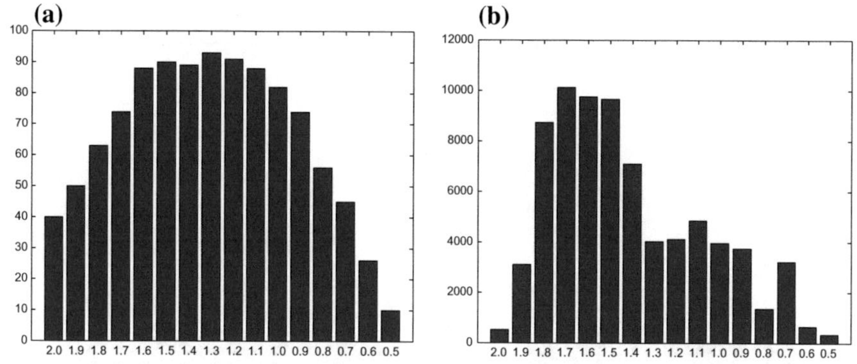

Fig. 4.14 Percentage ζ of successful runs of the $\mathrm{EP}_{N;\alpha}$ algorithms (**a**) and the mean number of iterations $\bar{t}$ needed for global extremum localization calculated over all successful runs (**b**) versus the stability index α for the 2D Rastringin function

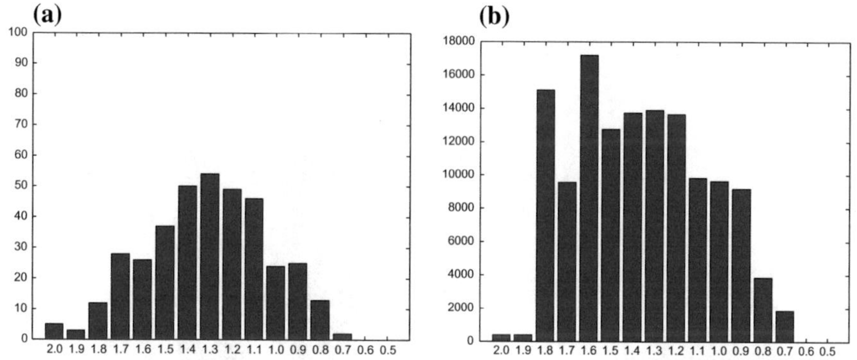

Fig. 4.15 Percentage ζ of successful runs of the $\mathrm{EP}_{N;\alpha}$ algorithms (**a**) and the mean number of iterations $\bar{t}$ needed for global extremum localization calculated over all successful runs (**b**) versus the stability index α for the 5D Rastringin function

the saddle crossing ability of $\mathrm{ESSS}_{N;2}$ with n increasing. All observations show that the range of the dead surrounding is very sensitive to the space dimension n, scale parameter σ and stability index α. Therefore, the problem of selection of the algorithm steering parameter needs additional study.

Like in the previously considered experiments, it can be noticed that the self-adaptation mechanism of the scale parameter in the $\mathrm{EP}_{N;\alpha}$ algorithms allows eliminating the dead surrounding effect. But this effect influences the effectiveness of algorithms of the $\mathrm{EP}_{N;\alpha}$ class in an indirect manner. In order to eliminate the dead surrounding effect, the scale parameter σ rapidly decreases to very low values and the $\mathrm{EP}_{N;\alpha}$ process is trapped around a local extremum. Thus, $\mathrm{EP}_{N;\alpha}$ algorithms with low values of α either find a global extremum very quickly or do not find it at all. In the case of large values of α, the effectiveness of $\mathrm{EP}_{N;\alpha}$ is low because the probability of macro-mutations is very low.

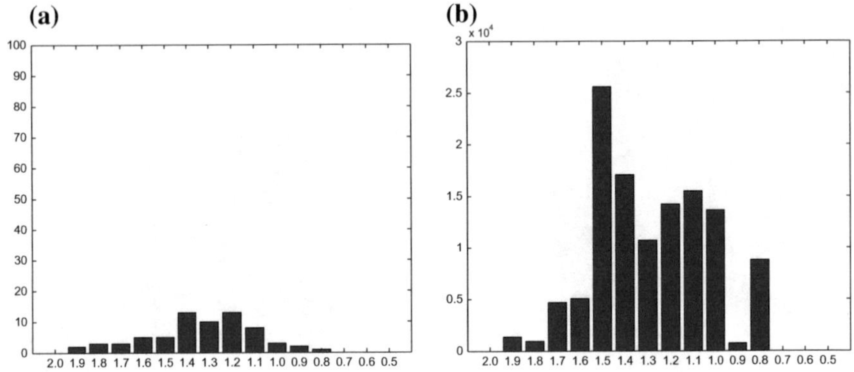

Fig. 4.16 Percentage ζ of successful runs of the $\mathrm{EP}_{N;\alpha}$ algorithms (**a**) and the mean number of iterations $\bar{t}$ needed for global extremum localization calculated over all successful runs (**b**) versus the stability index α for the 10D Rastringin function

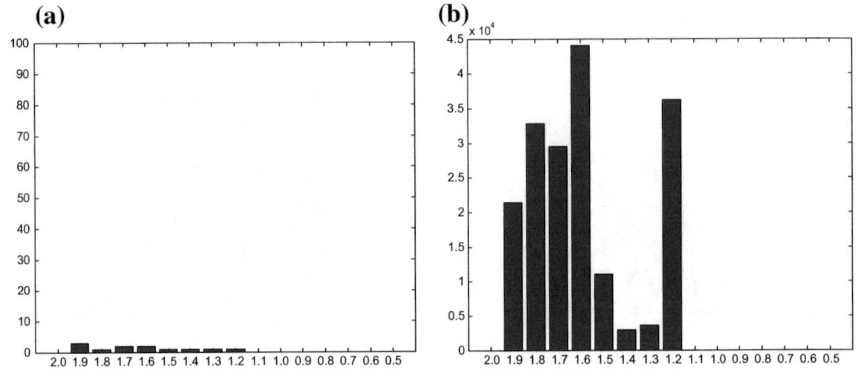

Fig. 4.17 Percentage ζ of successful runs of the $\mathrm{EP}_{N;\alpha}$ algorithms (**a**) and the mean number of iterations $\bar{t}$ needed for global extremum localization calculated over all successful runs (**b**) versus the stability index α for the 20D Rastringin function

4.2 Symmetry Effect

4.2.1 Local Convergence of $(1+1)ES_{N;\alpha}$

4.2.1.1 Problem Statement

Let us consider the modification $(1+1)ES_{N;\alpha}$ of the evolutionary strategy $(1+1)ES$ (Rechenberg 1965). The population in iteration t is reduced to only one element $\boldsymbol{x}^t$. The descendant $\boldsymbol{y}^t$ of the element $\boldsymbol{x}^t$ is obtained by the simple mutation operation

$$\boldsymbol{y}^t = \boldsymbol{x}^t + \sigma \boldsymbol{Z}, \tag{4.8}$$

where $\boldsymbol{Z} \sim S\alpha S$ for given α and σ is the input steering parameter of the algorithm. The better element, in the sense of fitness function values, of the pair $\boldsymbol{x}^t$ and $\boldsymbol{y}^t$ is chosen as a new base element $\boldsymbol{x}^{t+1}$ for further search, i.e.,

$$\boldsymbol{x}^{t+1} = \begin{cases} \boldsymbol{x}^t \text{ if } \Phi(\boldsymbol{x}^t) > \Phi(\boldsymbol{y}^t), \\ \boldsymbol{y}^t \text{ for other cases.} \end{cases} \tag{4.9}$$

Substituting $t \leftarrow t + 1$, the operations (4.8) and (4.9) are iteratively repeated until a given stop criterion is met.

The aim of this experiment is the analysis of the $(1 + 1)ES_{N;\alpha}$ algorithm's exploitation ability. The algorithm $(1 + 1)ES_{N;\alpha}$ is repeatedly started from different points. Results for different initial points are compared using the percentage ζ of successful mutations, i.e., those that give a better descendant than the basic element in the sense of the fitness function.

4.2.1.2 Experiment and Result Analysis

The four dimensional sphere function (B.1) is used in the simulation experiment. Four initial points are chosen:
$\boldsymbol{a}_1 = (100, 0, 0, 0)$,
$\boldsymbol{a}_2 = (100/\sqrt{2}, 100/\sqrt{2}, 0, 0)$,
$\boldsymbol{a}_3 = (100/\sqrt{3}, 100/\sqrt{3}, 100/\sqrt{3}, 0)$,
$\boldsymbol{a}_4 = (50, 50, 50, 50)$.

It is worth noting that $\|\boldsymbol{a}_i\| = 100$ for each $i = 1, 2, 3, 4$. Four algorithms of the class $(1 + 1)ES_{N;\alpha}$ are used in simulations: $(1 + 1)ES_{N;2}$, $(1 + 1)ES_{N;1.5}$, $(1 + 1)ES_{N;1}$ and $(1 + 1)ES_{N;0.5}$. The scale parameter is chosen the same for all algorithms: $\sigma = 0.1$. Each algorithm was started 500 times for each initial point.

The percentage of successful mutations for all algorithms considered and all initial points is presented in Fig. 4.18. Let $\zeta_{\alpha,i}$ represent the percentage of successful mutations of the algorithm $(1 + 1)ES_{N;\alpha}$ ($\alpha = 2, 1.5, 1, 0.5$), which was started from the point $\boldsymbol{a}_i$ ($i = 1, 2, 3, 4$).

It is easy to see that $\zeta_{2;i}$ is independent of initial point selection. This independence is caused by the spherical symmetry of the normal distribution (Fig. 4.18). The value $\zeta_{2;i} \cong 50\%$ for each $i = 1, 2, 3, 4$ is caused by the fact that the mean range of Gaussian mutation (described by $\sigma = 0.1$) is significantly lower than the curvature of the contour of the function f_{sph}.

However, when α decreases, then $\zeta_{\alpha;i}$ rapidly decreases and disproportion between results for different initial points grows. The percentage of successful mutations $\zeta_{\alpha;i}$ decreases when the distance between the initial point and the axis of the reference frame increases. The obtained effect is caused by the symmetry of the distributions $NS\alpha S(\boldsymbol{\sigma})$, which prefers directions parallel to the axis of the reference frame (Fig. 3.3). Thus, in the case of $\boldsymbol{a}_1$, which is located on the axis of the reference frame, the probability of descendant location much closer to the optimum point of

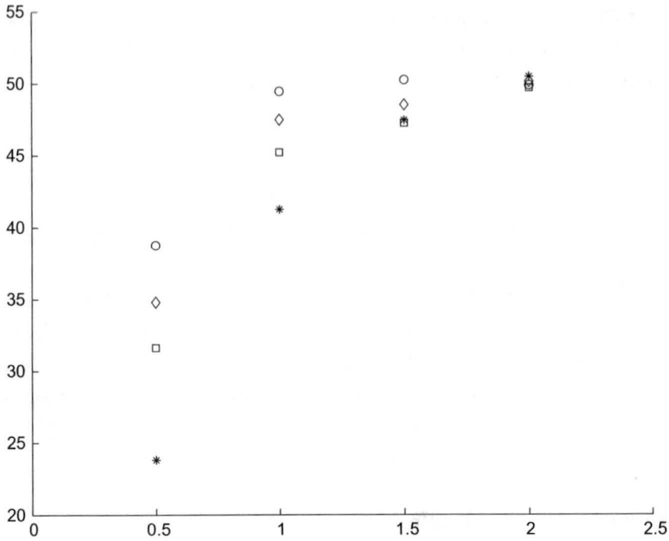

Fig. 4.18 Percentage ζ of successful mutations for $(1+1)ES_{N,\alpha}$ started from different points ($\boldsymbol{a}_1$: circles, $\boldsymbol{a}_2$: diamonds, $\boldsymbol{a}_3$: squares, $\boldsymbol{a}_4$: stars) versus α

the function f_{sph} is highest. This probability rapidly decreases with the number of non-zero coordinations of the initial point. The disproportion between results of the same strategy $(1+1)ES_{N;\alpha}$ but for different initial points increases with the stability index α decreasing.

Let $\bar{t}_{\alpha;i}$ describe the mean number of iterations needed for extremum localization (the stop criterion: $f_{sph} < 0.5$) calculated over 500 runs of the $(1+1)ES_{N;\alpha}$ algorithm ($\alpha = 2, 1.5, 1, 0.5$) started from the initial point $\boldsymbol{a}_i$ ($i = 1, 2, 3, 4$) (Fig. 4.19). It is surprising that, unlike in the case of the relation between $\zeta_{\alpha;i}$ and α, and $\boldsymbol{a}_i$, both $\bar{t}_{2;i}$ and $\bar{t}_{0.5;i}$ seem to be independent of initial point selection. Moreover, it can be expected that $\bar{t}_{0.5;4} < \bar{t}_{0.5;1}$ (when $\zeta_{0.5;4} \ll \zeta_{0.5;1}$; see Fig. 4.18); however, the difference between both values is of the same order as the statistical error. The strongest dependence between $\bar{t}_{\alpha;i}$ and $\boldsymbol{a}_i$ is obtained for $\alpha = 1.5$. The above observation suggests that there are two rival mechanisms influencing $\bar{t}_{\alpha;i}$. One is the relation (described above) between initial point selection and $\zeta_{\alpha;i}$. The other is connected with increasing tails' heaviness of symetric α-stable distributions $S\alpha S(\sigma)$ with α decreasing. For low values of the stability index α, on the one hand, most mutations are unsuccessful (the descendant is of worse quality than the parent), but on the other, the mean 'jump' of successful mutations is significantly larger than in the case of mutations with a larger value of α.

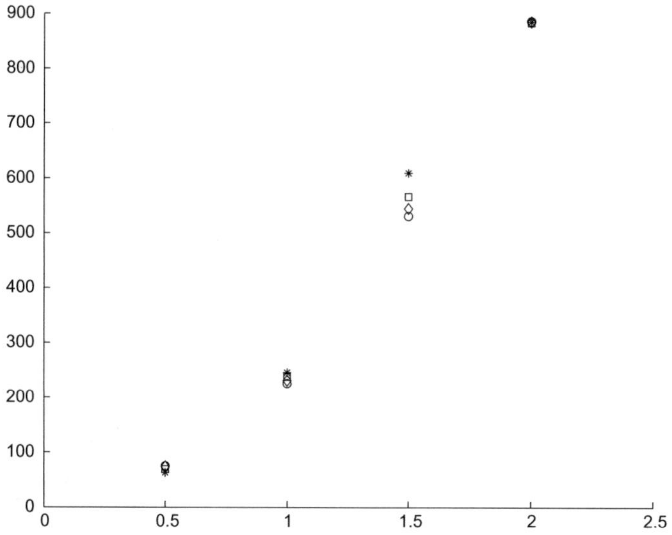

Fig. 4.19 Mean number of iterations needed for extremum neighborhood localization ($f_{sph} < 0.5$) for $(1+1)ES_{N;\alpha}$ started from different initial points ($\boldsymbol{a}_1$: circles, $\boldsymbol{a}_2$: diamonds, $\boldsymbol{a}_3$: squares, $\boldsymbol{a}_4$: stars) versus α

4.2.2 Saddle Crossing

Let us consider the following set of 4D fitness functions:

$$\Phi_l(\boldsymbol{x}) = \frac{1}{2}\exp\left(-5\|\boldsymbol{x}\|^2\right) + \exp\left(-5\|\boldsymbol{x}-\boldsymbol{m}_l\|^2\right), \quad l = 1, 2, 3, 4, \tag{4.10}$$

where
$\boldsymbol{m}_1 = (1, 0, 0, 0)$,
$\boldsymbol{m}_2 = (1/\sqrt{2}, 1/\sqrt{2}, 0, 0)$,
$\boldsymbol{m}_3 = (1/\sqrt{3}, 1/\sqrt{3}, 1/\sqrt{3}, 0)$,
$\boldsymbol{m}_4 = (1/2, 1/2, 1/2, 1/2)$
represent localizations of global extrema of the functions $\{\Phi_l(\boldsymbol{x}) | l = 1, 2, 3, 4\}$. The lower local extremum is located at the beginning of the reference frame. It is easy to notice that the distance between the local and the global extremum is the same for all functions Φ_l and is equal to unity. Initial populations for simulations described below are generated by η mutations of the initial point located in the local extremum $\boldsymbol{x}_0^0 = (0, 0, 0, 0)$. The aim of this experiment is measurement of the effectiveness of saddle crossing between the quality hills. It will be done by calculating the mean number of iterations $\bar{t}$ needed for global extremum neighborhood localization,

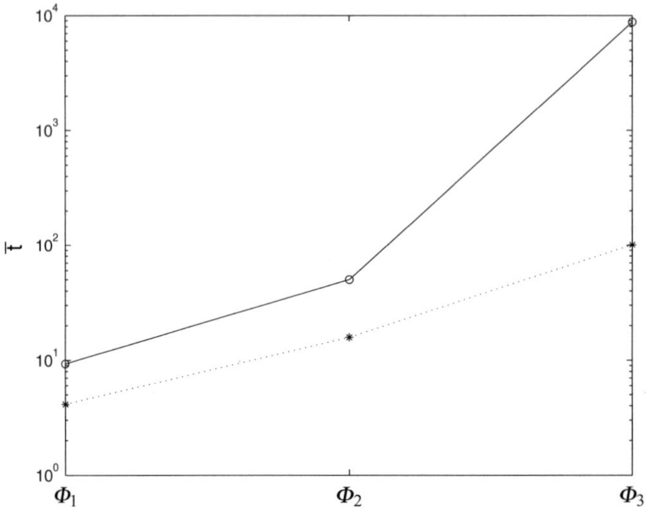

Fig. 4.20 Mean number of iterations needed for saddle crossing versus global extremum localization ($\mathrm{ESSS}_{N;1}$: circles, $\mathrm{ESSS}_{N;0.5}$: stars)

which means that the process is stopped when the population centre is located over the local extremum:

$$\exists k \quad \Phi_{sc}(\boldsymbol{x}_k) > \phi_{lim} > \Phi_{sc}(\boldsymbol{x}_0^0), \tag{4.11}$$

or $t_{\max}$ iterations is reached.

4.2.2.1 Evolutionary Search with Soft Selection $\mathrm{ESSS}_{N;\alpha}$

Two algorithms of the class $\mathrm{ESSS}_{N;\alpha}$: $\mathrm{ESSS}_{N;1}$ and $\mathrm{ESSS}_{0.5}$ are used. The parameters: the population size $\eta = 20$, the scale parameter $\sigma = 0.02$, the maximum number of iterations $t_{\max} = 10^5$, the limit fitness $\phi_{lim} = 0.6$ are given the same for all algorithms runs. Each algorithm was started 300 times for all sets of parameters. Results for the three functions $\{\Phi_l(\boldsymbol{x})|l = 1, 2, 3\}$ (4.10) and algorithms based on the $NS\alpha S(\boldsymbol{\sigma})$ mutation with the lowest stability indices are presented in Fig. 4.20. The relation between saddle crossing effectiveness of the evolutionary processes considered and global extremum localization is clearly seen.

4.2.2.2 Evolutionary Programming $\mathrm{EP}_{N;\alpha}$

Next, 16 algorithms of the class $\mathrm{EP}_{N;\alpha}$ ($\alpha = 2.0, 1.9, 1.8, \ldots, 0.6, 0.5$) were tested in saddle crossing of the four functions $\Phi_l \mid l = 1, 2, 3, 4$. The initial population of $\eta = 20$ individuals is randomly chosen with a uniform distribution from the sets

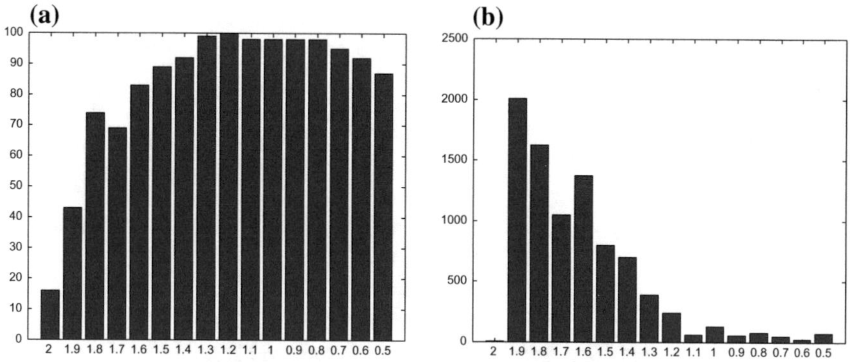

Fig. 4.21 Percentage of successful runs ζ of the $\mathrm{EP}_{N;\alpha}$ algorithm (**a**) and the mean number of iterations $\bar{t}$ needed for saddle crossing taken over all successful algorithm runs (**b**) versus the stability index α for $\Phi_1(\boldsymbol{x})$ (4.10)

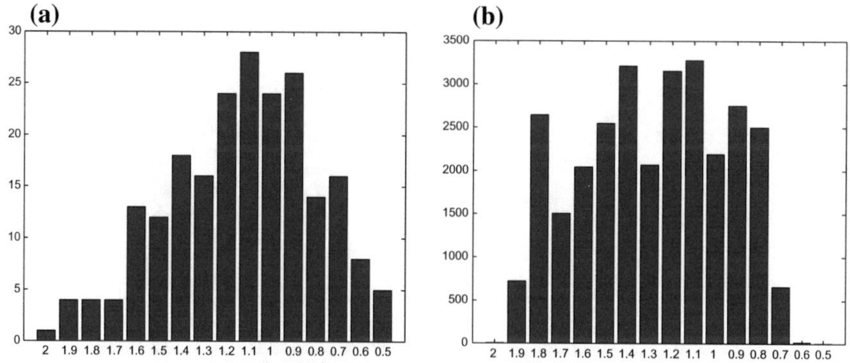

Fig. 4.22 Percentage of successful runs ζ of the $\mathrm{EP}_{N;\alpha}$ algorithm (**a**) and the mean number of iterations $\bar{t}$ needed for saddle crossing taken over all successful algorithm runs (**b**) versus the stability index α for $\Phi_2(\boldsymbol{x})$ (4.10)

$\Omega_x = \prod_{i=1}^{n}[-0.05, 0.05]$ and $\Omega_\sigma = \prod_{i=1}^{n}[0, 0.05]$, the number of selection sparing partners $q = 5$, the maximal number of iterations $t_{\max} = 10000$, the limit fitness $\phi_{lim} = 0.6$. Each algorithm was started 100 times for each function. The results are shown in Figs. 4.21, 4.22, 4.23 and 4.24.

4.2.3 Symmetry Effect Versus Global Extremum Searching

In order to illustrate the influence of the lack of the spherical symmetry of mutation based on the $NS\alpha S(\boldsymbol{\sigma})$ distribution on the effectiveness of global optimum searching in the multimodal environment, results described by Prętki (2008) will be presented below. Two 4D ($n = 4$) test functions were used in this experiment: the generalized

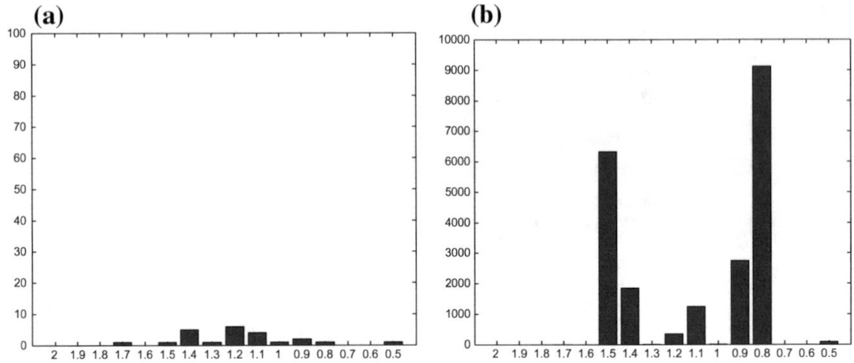

Fig. 4.23 Percentage of successful runs ζ of the $\text{EP}_{N;\alpha}$ algorithm (**a**) and the mean number of iterations $\bar{t}$ needed for saddle crossing taken over all successful algorithm runs (**b**) versus the stability index α for $\Phi_3(\boldsymbol{x})$ (4.10)

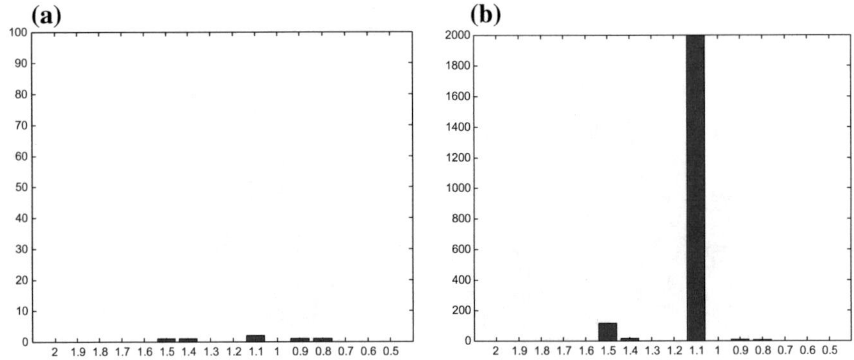

Fig. 4.24 Percentage of successful runs ζ of the $\text{EP}_{N;\alpha}$ algorithm (**a**) and the mean number of iterations $\bar{t}$ needed for saddle crossing taken over all successful algorithm runs (**b**) versus the stability index α for $\Phi_4(\boldsymbol{x})$ (4.10)

Ackley function $f_A(\boldsymbol{x})$ (B.3) and the Rastrigin function $f_R(\boldsymbol{x})$ (B.4). Both environments possess many local optima, which are located in nodes of 4D hypercubic nets, whose axes are parallel to those of the reference frame. In order to emphasize the influence of the symmetry effect on evolutionary search efficiency, rotations of the reference frame were done. These rotations are based on the following directional vectors:

$$\boldsymbol{b}_1 = [1, 0, 0, 0]^T,$$
$$\boldsymbol{b}_2 = [\frac{1}{\sqrt{2}}, \frac{1}{\sqrt{2}}, 0, 0]^T,$$
$$\boldsymbol{b}_3 = [\frac{1}{\sqrt{3}}, \frac{1}{\sqrt{3}}, \frac{1}{\sqrt{3}}, 0]^T,$$

$$b_4 = [0.5, 0.5, 0.5, 0.5]^T$$

and the respective rotation matrices:

$$M_1 = I_4,$$

$$M_k = [I_4 - 2v_k v_k^T], \quad v_k = \frac{b_1 - b_k}{\|b_1 - b_k\|}, \quad k = 2, 3, 4.$$

Test functions mentioned above were transformed according to the formula $f(M_k x)$, $k = 1, 2, 3, 4$. Algorithms of the class $\mathrm{ESTS}_{N;\alpha}$ are applied in the experiment. Mutation is based on the stable vector $X \sim NS\alpha S(\sigma)$ with the scale parameter σ, which was selected in order to guarantee algorithm convergence during $T_{\max} = 1000$ generations. Values of the other parameters were chosen as follows: the population size $\eta = 20$, the size of the tournament group $\eta_G = 4$, the starting point $x_0 = M_k[10, 10, 10, 10]^T$. Figures 4.25 and 4.26 present relations between the fit-

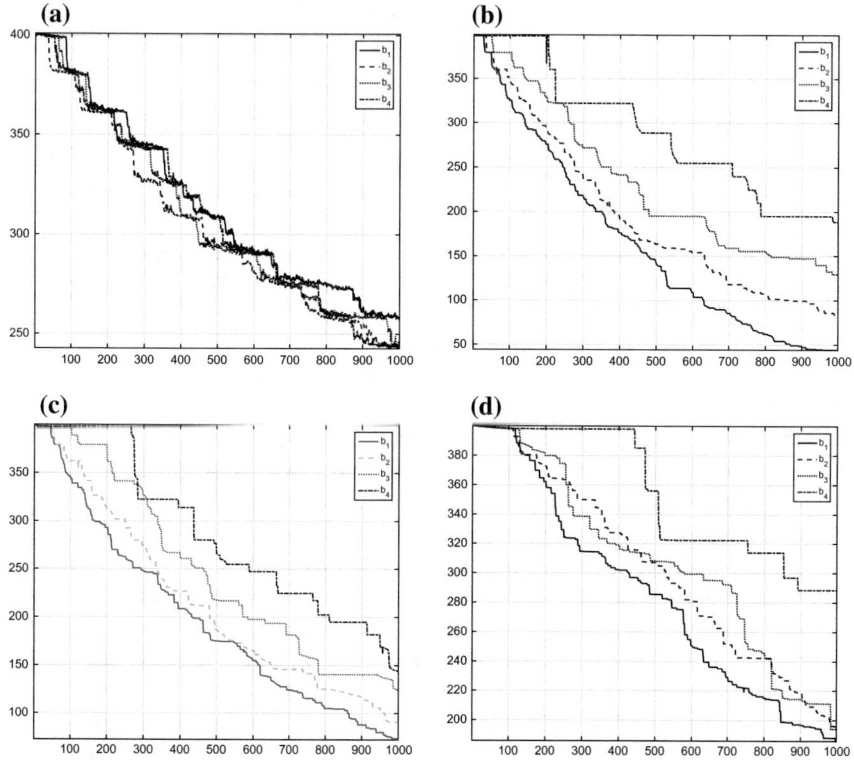

Fig. 4.25 Process of finding the best solution in population for Ackley's function ($n = 4$)—$\mathrm{ESTS}_{N;\alpha}$ algorithm with mutation based on the stable vector: $\alpha = 2$ (**a**), $\alpha = 1.5$ (**b**), $\alpha = 1$ (**c**), $\alpha = 0.5$ (**d**)

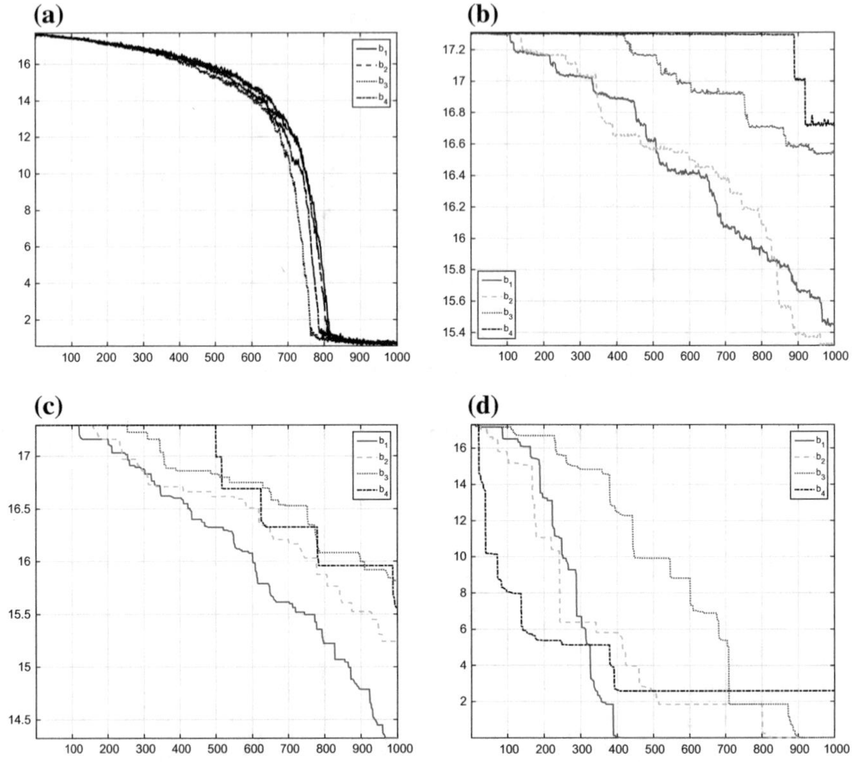

Fig. 4.26 Process of the best solution in the population for Rastrigin's function ($n = 4$)—ESTS$_{N;\alpha}$ algorithm with mutation based on the stable vector: $\alpha = 2$ (**a**), $\alpha = 1.5$ (**b**), $\alpha = 1$ (**c**), $\alpha = 0.5$ (**d**)

ness of the best element in the population versus iterations—results are averaged over 50 independent algorithm runs.

Orthogonal rotation matrices do not change objective functions, but the reciprocal location of profitable areas and directions preferred by mutation operations. It is easy to notice that even slight changes, represented by the vector $\boldsymbol{b}_2$, can cause a rapid decrease in evolutionary search efficiency. It is only in the case of Gaussian mutation ($\alpha = 2$), which has a spherical symmetry, that significant differences in the efficiency of the evolutionary algorithms for different vectors $\boldsymbol{b}_i$ are not observed.

The evolutionary process depends on the searching space rotation angle. When local extrema of testing functions are located in nodes of the net with the axis parallel to that of the reference frame ($\boldsymbol{b}_1$), the averaged trajectory of the best solution is characterized by significantly smaller leaps, unlike in the case of rotated spaces ($\boldsymbol{b}_2, \boldsymbol{b}_3, \boldsymbol{b}_4$). This means that the evolutionary process mainly consists in hurdling by a neighboring landscape saddles located in directions which are parallel to the axis of the reference frame—preferred by the mutation operation. Taking into account rotated spaces, especially in the case of $\boldsymbol{b}_4$, the averaged trajectory of the best solution consists of relatively long time intervals of a lack of progression, which are erupted by rapid jumps of objective function values—this indicates *successful* macro-mutation, which occurs so infrequently that a decrease in evolutionary process efficiency is noticeable.

4.3 Summary

Mutation based on the $NS\alpha S(\sigma)$ distribution is most popular in the literature. Unfortunately, it is burdened by two strong properties—the dead surrounding and symmetry effects. The former occurs in multidimensional spaces and is connected with the existence of some neighborhood around the mutated point with a very low probability of descendant location. The size of this surrounding increases quite quickly with the searching space size increasing as well as the stable index α decreasing. The simulation experiments presented in this chapter show strong influence of the dead surrounding effect on the process of local extremum localization. The $\mathrm{ESSS}_{N;\alpha}$ algorithm, which has no mechanism of scale parameter adaptation, is converged to some state of an selection–mutation equilibrium—the population is kept at a distance from an optimal point. This distance rapidly increases with the increase of the space dimension as well as of the scale parameter. If the scale parameter adaptation mechanism is included in an evolutionary process, e.g., $\mathrm{EP}_{N;\alpha}$ is applied, then the range of the dead surrounding can be reduced, but it is related to rapid decreasing of the scale parameter during the evolutionary process, and the population quickly gets stuck in a local extremum trap (sometimes in the global one).

The symmetry effect is connected with the lack of the spherical symmetry of the $NS\alpha S(\sigma)$ distribution with $\alpha < 2$ and a spectral measure Γ_s, whose probability mass is uniformly distributed in points -1 and 1 of each axis of the reference frame. Therefore, the $NS\alpha S(\sigma)$ distribution prefers directions parallel to axes of the reference frame. The strength of this preference increases with the stability index decreasing. The presented simulation experiments show strong influence of reference frame selection in the $\mathbb{R}^n$ space on global extremum localization efficiency of evolutionary algorithms with mutations based on the $NS\alpha S(\sigma)$ distribution. At the same time, it was shown that the influence of the symmetry effect on global extremum localization effectiveness (in the case of low α, where this influence is strongest) can be, ironically, partially reduced by high frequency of the occurrence of macro-mutations.

References

Beyer, H. G. (2001). *The theory of evolution strategies*. Heidelberg: Springer.
Karcz-Dulęba, I. (2004). Time to convergence of evolution in the space of population states. *International Journal Applied Mathematics and Computer Science*, *14*(3), 279–287.

Obuchowicz, A. (2001). On the true nature of the multi-dimensional Gaussian mutation. In V. Kurkova, N. C. Steel, R. Neruda, & M. Karny (Eds.), *Artificial neural networks and genetic algorithms* (pp. 248–251). Vienna: Springer.

Prętki, P. (2008). *α-Stable distributions in evolutionary algorithms for global parametric optimization*. Ph.D. thesis, University of Zielona Góra (in Polish).

Rechenberg, I. (1965). Cybernetic solution path of an experimental problem. *Royal Aircraft Establishment*, Library Translation 1122, Farnborough, Hants.

Shao, J. (1999). *Mathematical statistics*. New York: Springer.

Chapter 5
Isotropic Stable Mutation

Co-author: Przemysław Prętki

In the previous chapter, it was shown that application of the non-isotropic stable distributions $NS\alpha S(\boldsymbol{\sigma})$ (3.47) to the mutation operation in evolutionary algorithms working in $\mathbb{R}^n$ is connected with two of their properties, which strongly influence the decreasing of searching effectiveness of optimal solutions. These are the lack of the spherical symmetry of distributions with the stability index $\alpha < 2$ and the existence of the so-called dead surrounding. The symmetry effect can be removed if the non-isotropic distribution $NS\alpha S(\boldsymbol{\sigma})$ (3.47) is substituted by the isotropic one $IS\alpha S(\sigma)$ (3.64). The question is: What about the dead surrounding effect in this case? This problem is the subject of our discussion at the beginning of this chapter. Next, local convergence of evolutionary algorithms with isotropic stable mutation is analyzed and as attempt at determining a consensual balance between the exploration and exploitation abilities of these algorithms is made. The analysis of an evolutionary process with isotropic stable mutation robustness to the incorrectness of steering parameter selection (especially the scale parameter) and the possibility of application of the scale parameter self-adaptation mechanism are also crucial problems.

5.1 Dead Surrounding Effect

5.1.1 Probability Density Function of the Norm of the Isotropic Stable Vector

In accordance with the part 3 of Theorem 3.17, the isotropic stable vector $\boldsymbol{X} \sim IS\alpha S(\sigma)$ can be decomposed to the form

$$\boldsymbol{X} \stackrel{d}{=} \|\boldsymbol{X}\|\boldsymbol{u}^{(n)}. \tag{5.1}$$

A. Obuchowicz, *Stable Mutations for Evolutionary Algorithms*,
Studies in Computational Intelligence 797,
https://doi.org/10.1007/978-3-030-01548-0_5

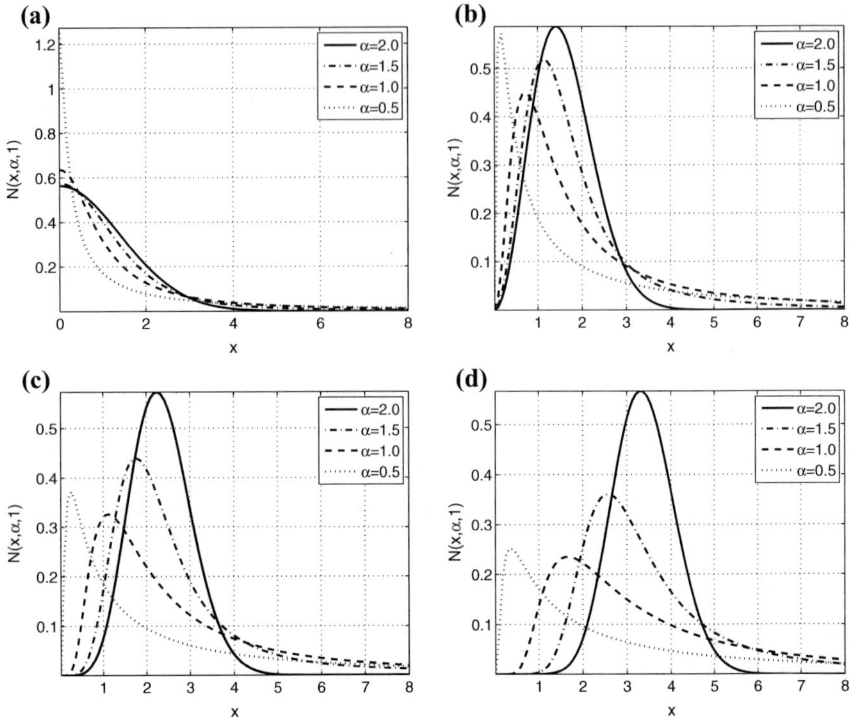

Fig. 5.1 Probability density function of the random variable $\|\boldsymbol{X}\|$, $\boldsymbol{X} \in \mathbb{R}^n$: $n = 1$ (**a**), $n = 3$ (**b**), $n = 6$ (**c**), $n = 12$ (**d**)

The dead surrounding analysis can be reduced to the analysis of the random variable $\|\boldsymbol{X}\|$. Unfortunately, the analytical form of this distribution in the general case is unknown. A quantitative numerical analysis can be based on the relation (3.75). In accordance with this relation, the probability density function $p(x)$ of the random variable $\|\boldsymbol{X}\|$ has the form

$$p(x) = c\frac{2\pi^{n/2}}{\Gamma(n/2)}x^{n-1}G(x^2), \tag{5.2}$$

where $G(\cdot)$ is the probability density generator of the stable random vector (3.69). The probability density functions (5.2) for four stability indices $\alpha = 2, 1.5, 1, 0.5$ and four dimensions $n = 1, 3, 6, 12$ of the search space are presented in Fig. 5.1.

Analyzing the probability density functions $p(x)$ (5.2) presented in Fig. 5.1, it can be observed that, like in the case of random vectors of a non-isotropic stable distribution, the dead surrounding also occurs in the case of $\|\boldsymbol{X}\|$ ($\boldsymbol{X} \sim IS\alpha S(\sigma)$). However, there exists some very important difference between distributions of norms of the non-isotropic stable random vector (3.47) and the isotropic one (3.64). In the former case (Fig. 4.1), the dead surrounding range increases with the stable index

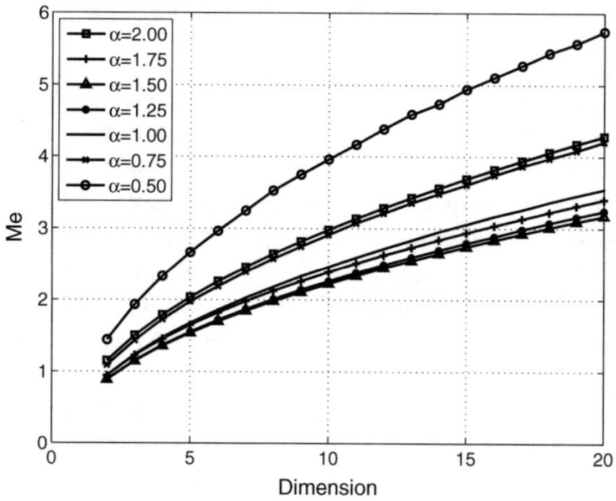

Fig. 5.2 Median of the random variable $\|\boldsymbol{X}\|$, $\boldsymbol{X} \sim IS\alpha S(\sigma)$

increasing. In the case of the isotropic stable random vector (Fig. 5.1), this relation is quite opposite—the maximum of the 'probability mass' is closer to the beginning of the reference frame for the lowest values of α. This observation allows presuming that the dead surrounding effect for heavy-tail distributions is less cumbersome than in the case of the normal distribution.

Distributions of norms of isotropic stable random vectors are unimodal (Fig. 5.1). Moreover, some order can be noticed: the lower the stability index α, the lower the distribution mode. An attempt at calculation of a median of the random variable $\|\boldsymbol{X}\|$ is presented by Prętki (2008). It can be reduced to the problem of local optimization:

$$\mathrm{Me}(\alpha, n) = \arg\min_{Me>0} \left[\int_0^{Me} p(h)dh - 1/2 \right]^2, \tag{5.3}$$

where $p(\cdot)$ is defined by (5.2). Figure 5.2 shows solutions of the problem (5.3).

Moreover, Prętki (2008) describes the following relations: the median versus the stability index and the search space dimension (with the relative error lower than 0.05),

$$\mathrm{Me}(\alpha, n) = (0.8\sqrt{n} - 0.35)\alpha^2 + (1.02 - 2.25\sqrt{n})\alpha + 2.29\sqrt{n} - 0.88. \tag{5.4}$$

The dependence between the median and the stability index is quite important. The minimum is obtained for

$$\alpha_{\min} \simeq \frac{2.25\sqrt{n} - 1.02}{1.6\sqrt{n} - 0.7}, \tag{5.5}$$

thus $\alpha_{\min} \simeq 1.37$ for $n = 1$ and $\alpha_{\min} \simeq 1.41$ for $n \to \infty$. One can suppose that the dead surrounding effect can have the smallest influence on local extremum localization efficiency for these values of the stability index.

5.1.2 Soft Selection Versus the Dead Surrounding

The influence of the dead surrounding effect on an evolutionary process is especially well exposed in the case of evolutionary algorithms with soft selection. In this case, the evolutionary process stabilizes in a selection–mutation counterpoise at some distance around the local extremum—the probability of descendant location in the neighborhood of this extremum is very low. Precise determination of the dead surrounding range is possible only in some simplest cases, e.g., for the population of two elements evolving in one-dimensional landscape with Gaussian mutation (Karcz-Dulęba 2004). In other cases, we can analyze it only using numerical simulations.

In order to do that, algorithms from the $\mathrm{ESTS}_{I;\alpha}$ class with mutation based on isotropic stable distributions and tournament selection will be applied. The function

$$f_{norm}(\boldsymbol{x}) = \|\boldsymbol{x}\| \tag{5.6}$$

is chosen as a fitness one. This choice allows us to interpret values of $f_{norm}(\boldsymbol{x})$ as a distance between a current solution and an optimal one. Calculations are done using the following set of parameters: stability indices $\alpha = \{2, 1.75, 1.5, 1.25, 1, 0.75, 0.5\}$, searching space dimensions $n = \{2, 3, 4, \ldots, 40\}$, the population size $\eta = 20$, the tournament group size $\eta_G = 4$. The initial population is obtained by η mutations of the optimal point $\boldsymbol{x}_0 = [0, 0, \ldots, 0]^T$ and, after stabilization of the population in a selection–mutation counterpoise, the fitness of the best element in noticed. The searching process was started 50 times for each pair $\{\alpha, n\}$, and the saved results were averaged. They are shown in Fig. 5.3.

The range of the dead surrounding area increases with the search space dimension n and stability index α increasing (Fig. 5.3). It is interesting that this increase is almost

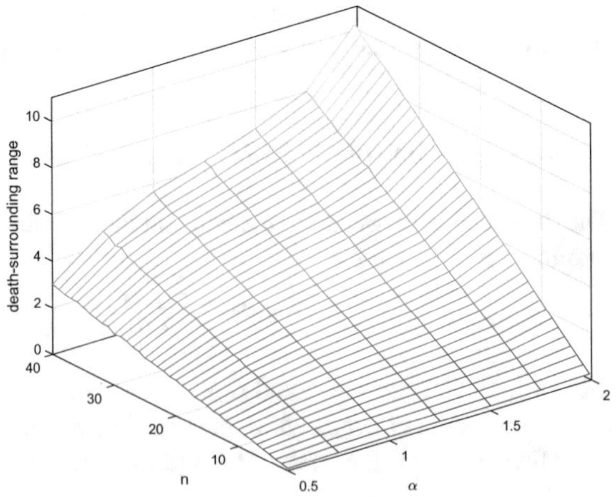

Fig. 5.3 Range of the dead surrounding area for the $\mathrm{ESTS}_{I;\alpha}$ algorithms

linear in regard to n for each α. The conclusion is that evolutionary algorithms with soft selection and isotropic stable mutation localize the extremum precisely in the case of low values of α.

5.2 Local Convergence Analysis

Local convergence of a stochastic optimization algorithm, especially an evolutionary one, can be described by the so-called *progress rate* φ, which is defined as an expectation value of a change of the distance between a current solution and an optimal one $\boldsymbol{x}^*$ in the following iterations:

$$\varphi = E\{\|\boldsymbol{x}_k - \boldsymbol{x}^*\| - \|\boldsymbol{x}_{k+1} - \boldsymbol{x}^*\|\}. \tag{5.7}$$

Progress rate analysis can give important knowledge about evolutionary process dynamics. For example, this type of analysis leads to a widely know regula of adaptation of the probability distribution of mutation: the *1/5 success regula* (Beyer and Schwefel 2002). Unfortunately, exact calculation of φ from the expression (5.7) is possible only in the case of some simple algorithms and simple objective functions (Auger and Hansen 2006; Beyer 2001; Bienvenue and Francois 2003; Rudolph 1996). Taking into account the above facts, we try to analyze the local convergence of the evolutionary strategy $(1+1)\mathrm{ES}_{I;\alpha}$ with isotropic stable mutation.

5.2.1 Progress Rate

Let us consider the following objective function:

$$f_{gsph}(\boldsymbol{x}) = a(\boldsymbol{x}^T\boldsymbol{x})^{b/2}, \tag{5.8}$$

where $a, b \geq 0$, and a class of simple evolutionary strategies $(1+1)\mathrm{ES}_{I;\alpha}$. The stability index α^* which guarantees the fastest convergence to local extremum is looked for.

Because of the elitist succession in the $(1+1)\mathrm{ES}_{I;\alpha}$ strategy, the solution in the $(k+1)$th iteration fulfils the following formula:

$$\boldsymbol{x}_{k+1} = \begin{cases} \boldsymbol{x}_k + \boldsymbol{Z}_k & \text{if} \quad f_{gsph}(\boldsymbol{x}_k + \boldsymbol{Z}_k) < f_{gsph}(\boldsymbol{x}_k) \\ \boldsymbol{x}_k & \text{if} \quad f_{gsph}(\boldsymbol{x}_k + \boldsymbol{Z}_k) \geq f_{gsph}(\boldsymbol{x}_k) \end{cases}, \tag{5.9}$$

where $\boldsymbol{Z}_k$ is stable random vector realization. Rudolph (1996) proposed the following definition of the progress rate for an elitist evolutionary strategy (5.9) and the objective function (5.8):

$$c = \frac{f(\boldsymbol{x}_{k+1})}{f(\boldsymbol{x}_k)}. \tag{5.10}$$

In the case of the function (5.8), the progress rate (5.10) has the form

$$c = \left(\frac{\boldsymbol{x}_{k+1}^T \boldsymbol{x}_{k+1}}{\boldsymbol{x}_k^T \boldsymbol{x}_k}\right)^{b/2}. \tag{5.11}$$

Taking into account that the optimal solution of the function (5.8) has the form $\boldsymbol{x}^* = \mathbf{0}$, as well as $\|\boldsymbol{x}\| = \left(\boldsymbol{x}^T \boldsymbol{x}\right)^{1/2}$, we can convert the relation (5.11) as follows:

$$c^{1/b} = \frac{\|\boldsymbol{x}_{k+1} - \boldsymbol{x}^*\|}{\|\boldsymbol{x}_k - \boldsymbol{x}^*\|}. \tag{5.12}$$

It is easy to see that, in the case of the spherical function ($b = 2$), the factor c is strongly connected with the definition of φ (5.7) using the following relation:

$$\varphi = \|\boldsymbol{x}_k\|(c - 1). \tag{5.13}$$

The selection operation (5.9) and the stochastic mutation $\boldsymbol{x}_{k+1} = \boldsymbol{x}_k + \boldsymbol{Z}_k$ are applied in the $(1+1)\mathrm{ES}_{I;\alpha}$ strategy, thus we introduce a new additional random value V in the form

$$V = \left(\frac{[\boldsymbol{x}_k + \boldsymbol{Z}]^T [\boldsymbol{x}_k + \boldsymbol{Z}]}{\boldsymbol{x}_k^T \boldsymbol{x}_k}\right)^{b/2}, \tag{5.14}$$

and the progress rate (5.10) can be also treated as a random value C of the form

$$C = \min\{V, 1\}. \tag{5.15}$$

The probability density function of the above random value does not exist, because all the probability mass for $P(V > 1) > 0$ is focused in one point $v = 1$.

Let us consider the expectation value of the progress rate (5.15):

$$\begin{aligned} E[C] &= E\Big[\min\Big\{V, 1\Big\}\Big] = \\ &= 1 + \int_0^1 v p_V(v; \alpha, \delta, b) dv - \int_0^1 p_V(v; \alpha, \delta, b) dv \\ &= 1 - \int_0^1 p_V(v; \alpha, \delta, b)[1 - v] dv, \end{aligned} \tag{5.16}$$

where $p_V(v, \alpha, \delta)$ is a probability density function of the random value V (5.14) and is of the form (Prętki 2008)

$$\begin{aligned} p_V(v;\alpha,\delta,b) = & \frac{2\delta^n v^{\frac{n-b}{b}} \pi^{\frac{n-1}{2}}}{b\Gamma(\frac{n-1}{2})} \times \\ & \times \int_{-1}^{1} G_\alpha\Big(\delta^2\big(v^{2/b} + 2\,v^{1/b}\,t + 1\big)\Big)(1-t^2)^{\frac{n-3}{2}}\,dt, \end{aligned} \tag{5.17}$$

where $\delta = \frac{\|x_k\|}{\sigma}$ is the length of the vector x_k measured in the scale parameter units and $G_\alpha(\cdot)$ is the probability density generator (3.69). If we take into account the objective function, then we can assume that isotropic stable mutation will be most effective in a local optimization process for possibly smallest values of $E[C]$ (5.16). Therefore, we are looking for an optimal value δ^* as a function the stability index α. We have to solve the following optimization problem (5.16):

$$\delta^*(\alpha) = \arg\max_{\delta>0} \int_0^1 p_V(v,\alpha,\delta)[1-v]dv. \tag{5.18}$$

Unfortunately, because of the complicated formulae of the probability density function $p_V(v;\alpha,\delta,b)$ (5.17), the problem (5.18) cannot be analytically solved. Numerical techniques have to be used instead.

5.2.2 *Convergence Analysis for n = 3*

Let us consider the three dimensional version of the optimization problem (5.18):

$$\delta^*(\alpha) = \arg\max_{\delta>0} \frac{2}{b}\delta^3\pi \int_0^1 \int_{-1}^1 [1-v]v^{\frac{3-b}{b}}\, G_\alpha\left(\delta^2\left(v^{2/b} + 2v^{1/b}t + 1\right)\right) dt dv. \tag{5.19}$$

The value of $\delta^*(\alpha)$ can be obtained by numerical optimization of the integral (5.19). Values of $\delta^*(\alpha)$ and the related values of the progress rate are presented in Tables 5.1 and 5.2. The relation between the expectation value $E[C(\delta)]$ and δ for the spherical function is illustrated in Fig. 5.4a.

Isotropic stable mutation applied to local optimization problems has a very interesting property. The evolutionary strategy $(1+1)\text{ES}_{I;\alpha}$ in its classical version, i.e., with mutation based on the Gaussian distribution, converges fastest to the local extremum (Fig. 5.4a). It is only in the nearest neighbourhood ($\|x - x^*\| \leqslant \sigma$) that distributions with low stability indices are more effective in extremum localization.

However, we must remember that the above property is obtained provided that the scale parameter of mutation $\sigma^* = (\|x - x^*\|)/\delta^*$ is correctly chosen (Table 5.1). Optimal selection of σ^* depends on our knowledge about the Euclidian distance between the current base solution of the strategy and the optimal one. In practice, we have no such knowledge. The results presented in Fig. 5.4 prove an interesting property connected with robustness of stable mutations to lack of knowledge about real distance $\|x - x^*\|$. If the course of $E[C(\delta)]$ is more oblate, like in the case of

Table 5.1 Quasi-optimal values of δ^*, which guarantee the fastest local convergence of the evolutionary strategy $(1+1)\text{ES}_{I;\alpha}$ to the optimal point of the 3D function $f_{gsph}(\boldsymbol{x})$ (5.8)

α	$b = 0.1$	$= 0.25$	$= 0.5$	$= 1.0$	$= 2.0$	$= 4.0$
2.00	2.28	2.30	2.35	2.43	2.61	2.96
1.75	2.29	2.32	2.38	2.49	2.69	3.08
1.50	2.33	2.37	2.44	2.56	2.78	3.24
1.25	2.40	2.44	2.51	2.65	2.90	3.37
1.00	2.49	2.58	2.62	2.78	3.07	3.54
0.75	2.68	2.67	2.80	2.97	3.26	3.64
0.50	3.21	2.89	3.37	3.50	3.56	4.20

Table 5.2 Progress rate $E(C(\delta^*))$ of the evolutionary strategy $(1+1)\text{ES}_{I;\alpha}$ and the 3D function $f_{gsph}(\boldsymbol{x})$ (5.8)

α	$b = 0.1$	$= 0.25$	$= 0.5$	$= 1.0$	$= 2.0$	$= 4.0$
2.00	0.9899	0.9757	0.9547	0.9200	0.8696	0.8076
1.75	0.9906	0.9775	0.9579	0.9253	0.8774	0.8171
1.50	0.9915	0.9795	0.9616	0.9314	0.8864	0.8287
1.25	0.9924	0.9818	0.9657	0.9384	0.8971	0.8431
1.00	0.9935	0.9843	0.9704	0.9466	0.9100	0.8612
0.75	0.9947	0.9873	0.9759	0.9563	0.9259	0.8842
0.50	0.9962	0.9908	0.9825	0.9681	0.9456	0.9141

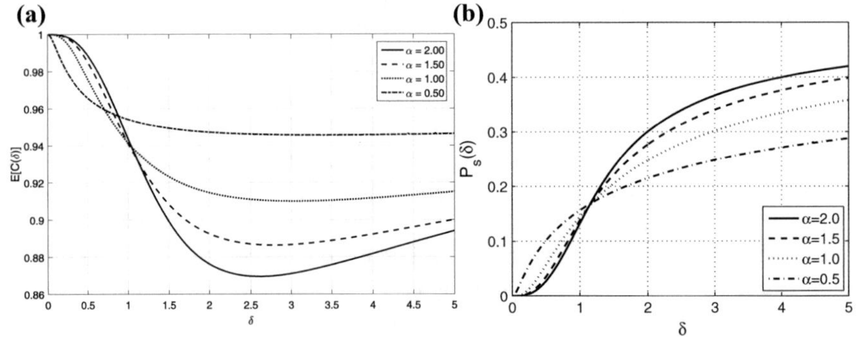

Fig. 5.4 Progress rate for the evolutionary strategy $(1+1)\text{ES}_{I;\alpha}$ and the 3D spherical objective function $f_{sph}(\boldsymbol{x})$ (**a**), as well as the related probability of success (**b**)

low stability indices, then the exploitation ability of the evolutionary process is less sensitive to the non-optimal choice of mutation parameters.

In the problem considered, the convergence of the evolutionary algorithm depends on correct selection of the scale parameter σ used in stable mutation. The optimal value σ^* is strongly connected with the distance between the current solution

Table 5.3 Probability of success $P_s(\alpha, \delta^*, b)$ calculated for the optimal values δ^* (Table 5.1)

α	$b = 0.1$	$= 0.25$	$= 0.5$	$= 1.0$	$= 2.0$	$= 4.0$
2.00	0.3250	0.3265	0.3302	0.3357	0.3470	0.3650
1.75	0.3139	0.3162	0.3204	0.3277	0.3397	0.3587
1.50	0.3029	0.3056	0.3103	0.3177	0.3299	0.3508
1.25	0.2880	0.2904	0.2945	0.3022	0.3146	0.3339
1.00	0.2780	0.2827	0.2847	0.2923	0.3047	0.3216
0.75	0.2648	0.2644	0.2695	0.2759	0.2857	0.2970
0.50	0.2537	0.2453	0.2576	0.2605	0.2619	0.2748

and the optimal one. This distance is usually unknown in practice, but it can be estimated during the evolutionary process. One of the oldest heuristics corresponding to Gaussian mutation, called the 1/5 regula (Beyer and Schwefel 2002), uses a specific estimator—the probability of success, i.e., the probability of successful mutation, where a descendant is better fitted than a parent. Knowing optimal values of the parameters presented in Table 5.1, optimal probabilities of success can be set:

$$P_s(\alpha, \delta, b) = \int_0^1 p_V(v; \alpha, \delta, b) dv, \tag{5.20}$$

where $p_V(v; \alpha, \delta, b)$ can be calculated from (5.17). In the 3D case ($n = 3$), (5.20) has the form

$$\begin{aligned} &P_s(\alpha, \delta, b) = \\ &= \tfrac{2}{b}\delta^3\pi \int_0^1 \int_{-1}^1 v^{\frac{3-b}{b}}\, G_\alpha\Big(\delta^2\big(v^{2/b} + 2v^{1/b}t + 1\big)\Big) dt dv. \end{aligned} \tag{5.21}$$

The values $P_s(\alpha, \delta, b)$ (5.21) for optimally configured distributions are presented in Table 5.3. The probability of success of the evolutionary strategy $(1+1)\text{ES}_{I;\alpha}$ with optimally configured stable mutation decreases with the stable index α decreasing. This proves that the 1/5 regula is not in effect for the whole class of stable mutations.

Local convergence analysis played an important role in deterministic optimization algorithms. It allows estimating time needed for optimal point localization with a given accuracy. In the case of evolutionary algorithms, such analysis allows analytical estimation of the influence of the given exploration distribution on algorithm efficiency of local extremum localization. In this subsection results of analysis of the evolutionary strategies $(1+1)\text{ES}_{I;\alpha}$ are presented. They clearly show the advantage of the normal distribution in local optimization tasks. This has been expected and is compatible with the results reported by Hansen et al. (2006) and Rudoph (1997). However, the above conclusion is true if steering parameters of mutation are correctly chosen. In practice, such a choice is usually impossible. So, the question about algorithm robustness to the choice of non-optimal parameters is justified. In the case of evolutionary strategies of the $(1+1)\text{ES}_{I;\alpha}$ class with mutations based on the $IS\alpha S(\sigma)$ distributions, strategies with low values of the stability index α are more robust. In the case of the normal distribution $\alpha = 2$, relatively small changes of algorithm steering parameters lead to a large decrease in its efficacy. This hypothesis is justified by the results presented in Fig. 5.4.

5.3 Exploration Versus Exploitation

The results presented so far in this chapter lead to the question about the possibility of achieving the exploration–exploitation balance for evolutionary algorithms with mutations based on the $IS\alpha S(\sigma)$ distributions. On the one hand, the analysis of the dead surrounding effect suggests that the most interesting distributions are these with the stability index $\alpha \simeq 1.4$ (5.5) or lower (Fig. 5.3), for which the local extremum neighborhood is closely exploited. On the other hand, macro-mutations are more probable for the above distributions because of their heavy tails, and thus the exploration ability also increases. The importance of the above properties is stressed by the limitation of the floating point representation because of its finite number of bits. This fact is the major cause of convergence dilution of many popular evolutionary algorithms, for which proofs of stochastic convergence to a global extremum exist. It is also the cause of the limitation of pseudo-random sequences. For example, it is practically impossible to obtain pseudo-random values of the normal distribution greater than 5σ; however, the theory allows obtaining any real value. In order to analyse the influence of the $IS\alpha S(\sigma)$ distribution on evolutionary algorithms exploration and exploitation abilities, two very simple numerical experiments are presented in this section.

5.3.1 *Local Convergence of the* $(1+1)ES_{I;\alpha}$ *Strategy*

The influence of α-stable distributions on the local convergence of evolutionary algorithms is illustrated by the following simulation experiments.

The n-dimensional spherical function $f_{sph}(\boldsymbol{x})$ (B.1) is again chosen as a fitness. One let the starting points have the form $\boldsymbol{x}_0 = [x_{0i} = \frac{100}{\sqrt{n}} \mid i = 1, \ldots, n]$, so that the distance between $\boldsymbol{x}_0$ and the extremum located at the beginning of the reference is equal to 100 independently of the search space dimension n. This problem is solved by a series of strategies $(1+1)\mathrm{ES}_{I;\alpha}$, i.e., those with a mutation operator based on the $IS\alpha S(\sigma)$ distribution. Let the space of parameters $(0, 2] \times \mathbb{R}_+$ of the stable distributions considered be discretized in the following way: $\alpha = 2.0, 1.5, 1.0, 0.5$, as well as $\sigma = 0.001, 0.002, \ldots, 0.009, 0.01, 0.02, \ldots, 0.09, 0.1, 0.2, \ldots, 0.9, 1, 2, \ldots, 10$. Algorithms are stopped if the best solution is not corrected during 100 iterations. Because our algorithms are stochastic, $N_t = 100$ independent $(1+1)\mathrm{ES}_{I;\alpha}$ processes for each parameter pair from the set $\{(\alpha_i, \sigma_j) \mid i = 1, \ldots, 4,\ j = 1, \ldots, 37\}$ and for different space dimensions $n = 1, 3, 6, 12$ are performed. So, this experiment is composed of 59200 independent algorithm runs.

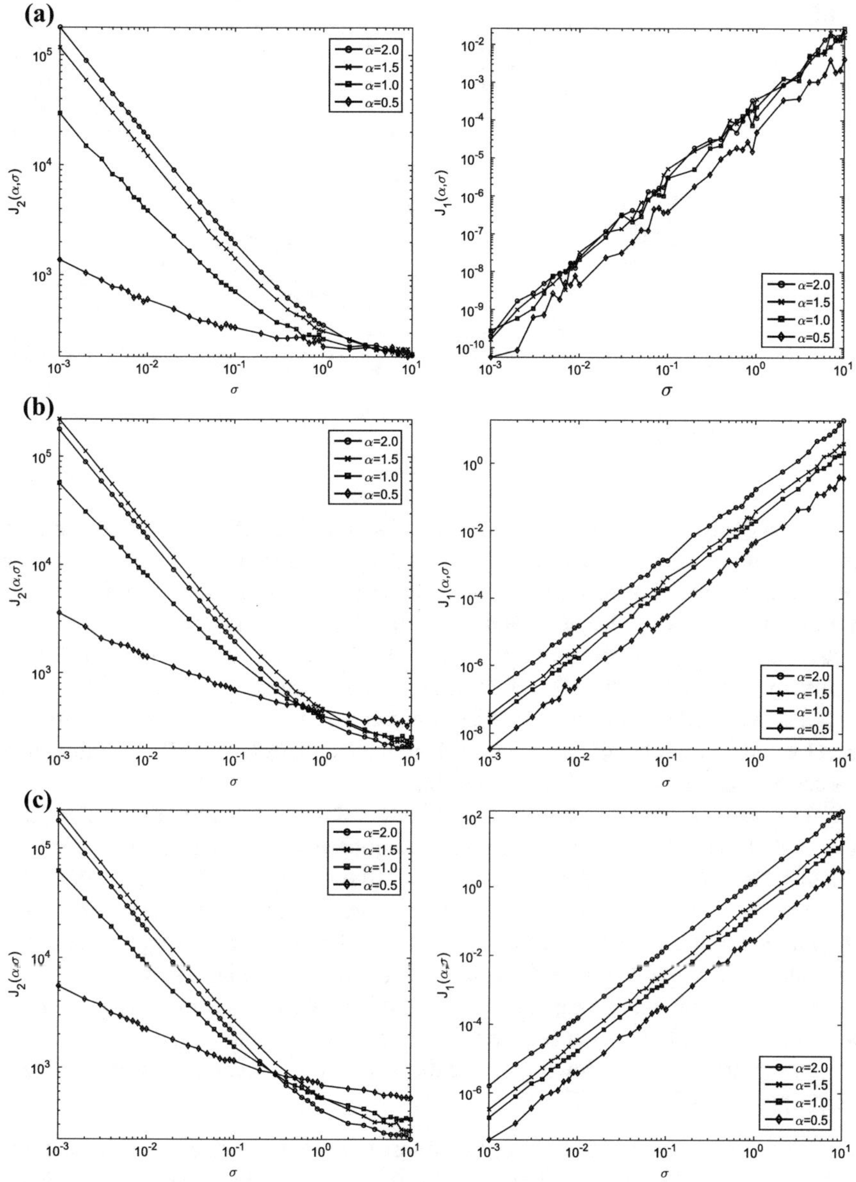

Fig. 5.5 Spherical function optimization using the $(1+1)\mathrm{ES}_{I;\alpha}$ strategies: left—the mean number of iterations needed for extremum localization; right—the mean value of the best solutions obtained by the strategy; **a**–**d** correspond to space dimensions $n = 1, 3, 6, 12$, respectively

The mean number of iterations needed for extremum localization and the mean value of the best solutions obtained by the evolutionary strategy over algorithm runs are presented in Fig. 5.5.

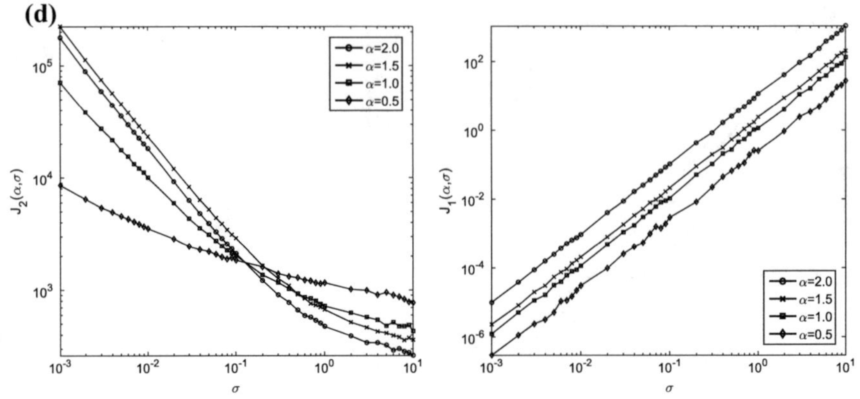

Fig. 5.5 (continued)

Let us analyze the charts on the left-hand side in Fig. 5.5. It is easy to see that, for a given scale parameter σ and space dimension n, mean extremum localization precision is better for evolutionary algorithms with smaller stability indices α. If we assume that, for the optimization task considered, the solution $\boldsymbol{x}^*$ is obtained with satisfactory precision when $f_{sph}(\boldsymbol{x}^*) < 0.01$, then the same solution quality is generally achieved for $\sigma < 0.1$. Taking into account charts on the right-hand side, one can see that the mean number of iterations needed to achieve the results obtained by $(1+1)\text{ES}_{I;0.5}$ is significantly lower than in the case of $(1+1)\text{ES}_{I;\alpha}$ with higher values of α and increases with α increasing.

The above regularity is diverted for higher values of the scale parameter. This means that the time needed for extremum neighborhood localization is shortest for the highest value of the stability index $\alpha = 2$ (the normal distribution case). At the same time, the precision of extremum localization decreases. It is interesting that the interval of scale parameter values, in which the order of convergence speed is inverted, moves to lower σ values with the space dimension increasing.

It is worth noting that the quality of the solutions decreases with the space dimension increasing. This is a consequence of the dead surrounding effect.

Let us introduce the following two quality criterions in order to analyse the exploration property of $(1+1)\text{ES}_{I;\alpha}$:

- $J_1(\alpha, \sigma)$: the solution quality, i.e., the fitness of the best obtained result which was found by the strategy $(1+1)\text{ES}_{I;\alpha}$ with mutation based on a stable distribution with the parameters (α, σ),
- $J_2(\alpha, \sigma)$: the number of iterations needed to achieve this solution for the same configuration.

Having both criterions, the Pareto optimality analysis can be carried out. The relation between the criterions is illustrated in Fig. 5.6.

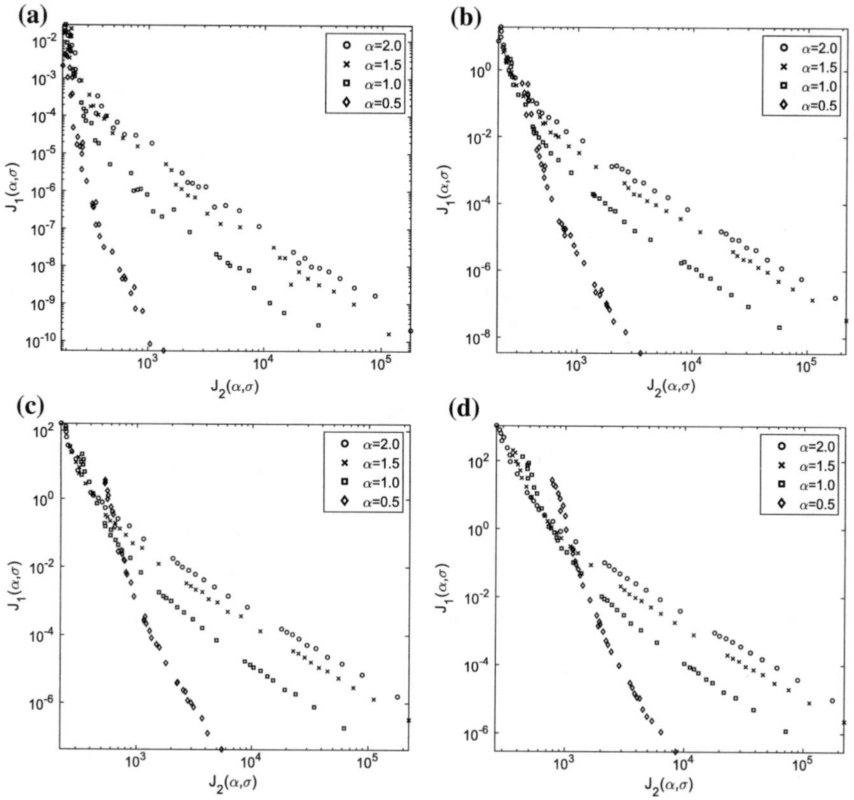

Fig. 5.6 Relation between mean values of $J_1(\alpha, \sigma)$ and $J_2(\alpha, \sigma)$ for the evolutionary strategies $(1+1)\mathrm{ES}_{I;\alpha}$. Points from left to right correspond to declining scale parameters σ: $n = 1$ (**a**), $n = 3$ (**b**), $n = 6$ (**c**) and $n = 12$ (**d**)

Analyzing the charts in Fig. 5.6 it can be observed that distributions with heavier tails have a tendency to dominate over distributions with higher stability indices α. Most of the Pareto front is created by solutions corresponding to $\alpha = 0.5$, and this part increases with the space dimension decreasing.

5.3.2 Saddle Crossing: The $(1, 2)ES_{I;\alpha}$ Strategy

The simple evolutionary strategy $(1, 2)\mathrm{ES}_{I;\alpha}$ is applied to the saddle crossing problem (Appendix A.1) in this subsection. This choice is based on the assumption that exploration properties are usually identical to an algorithm's abilities of saddle crossing between two quality peaks of an objective function, because this ability allows broadening a search area in the multidimensional real space on which the fitness function is defined (Galar 1990). The limit fitness (A.2) is chosen as $\phi_{lim} = 0.55$,

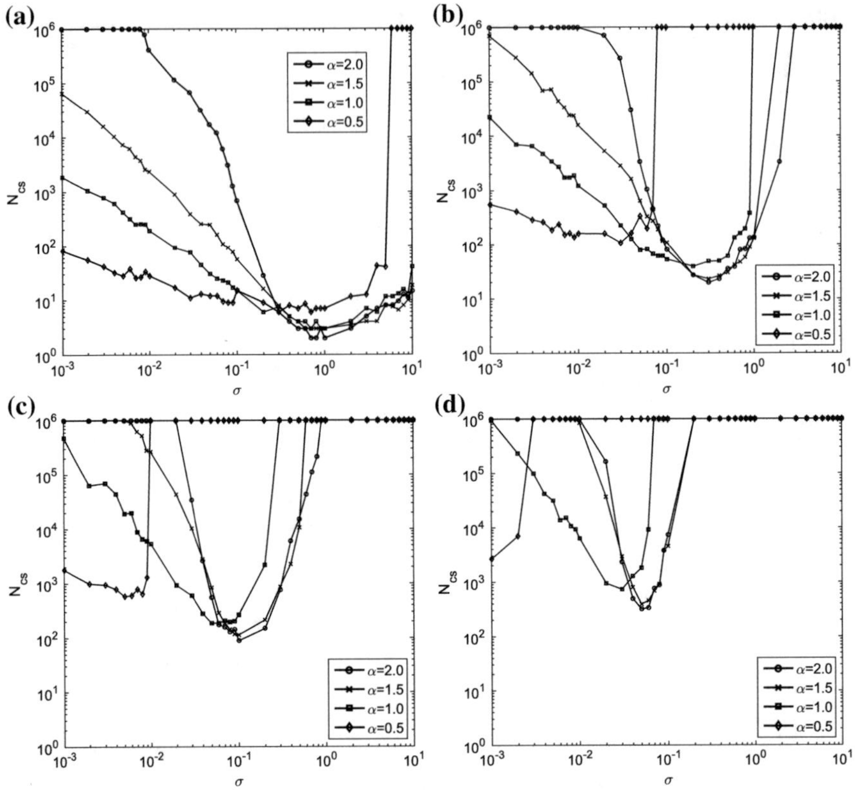

Fig. 5.7 Number of iterations N_{cs} needed to cross the saddle between two Gaussian peaks for different search space dimensions: $n = 1$ (**a**), $n = 3$ (**b**), $n = 6$ (**c**), $n = 12$ (**d**)

the maximal number of iterations $t_{\max} = 10^6$. The space of parameters of isotropic stable distributions is discretized identically as in the experiment described in the previous subsection. Mean numbers of iterations needed to cross the saddle obtained for a hundred independent runs of the $(1, 2)\mathrm{ES}_{I;\alpha}$ algorithm for each set of parameters $\{\alpha_i, \sigma_j\}$ are presented in Fig. 5.7.

Analyzing the results presented in Fig. 5.7, it is worth noting that all distributions considered in the experiment can guarantee similar saddle crossing effectiveness if only the scale parameter σ is properly chosen. However, this effect is very sensitive to the above choice of σ. The evolutionary window width formed by σ values, for which the $(1, 2)\mathrm{ES}_{I;\alpha}$ strategy crosses the saddle in less then $t_{\max} = 10^6$ iterations, decreases with α increasing. Thus, we can conclude that distributions with low stability indices are more robust to an imprecise choice of the scale parameter.

Observing the influence of the space dimension on evolutionary strategy effectiveness, it is easy to see that the evolutionary window for all distributions narrows down. This effect is one of the symptoms of the so-called curse of dimensionality, which is unavoidable in the case of isotropic distributions.

5.3.3 Conclusions

The normal distribution commonly applied to the mutation operator of evolutionary algorithms based on the floating point representation of individuals is the cause of quick convergence to a local extremum and allows quick crossing of evolutionary saddles. Unfortunately, both situations need a precise choice of the scale parameter σ. What is worse, both need quite different values of σ.

Based on two simple experiments: optimization of the spherical function (which is the basic model for effectiveness analyzes of local optimization methods) and the crossing of the saddle between two Gaussian peaks (commonly applied for effectiveness analysis of global optimization algorithms), one can point out that heavy-tail distributions possess some advantage. In the case of local optimization, distributions with the low value of the stability index dominate over others—they converge quickly and guarantee solutions of better quality. The presented results suggest that the commonly used normal distribution in the mutation operator of evolutionary algorithms should be substituted by stable distributions with low α, especially in the case of low space dimensions. In the case of high dimensions of the objective function domain, the exploitation ability of evolutionary algorithms with stable mutation based on heavy tail distributions decreases. This fact is connected with intensification of the dead surrounding effect (Fig. 5.2).

5.4 Robustness Analysis of Stable Mutation

The hypothesis that distributions with lower stability indices are more robust to an imprecise choice of the scale parameter was proposed at the end of Sect. 5.2. Now, we will try to verify this hypothesis. Namely, we assume that the optimized function belongs to the class of functions of positive continuous values from some n dimensional hyper-cube. Evolutionary strategies $(1+1)\mathrm{ES}_{I;\alpha}$ are used to optimize this function. We pose a statistical thesis: for a randomly chosen scale parameter of isotropic stable mutation, the probability of an effective $(1+1)\mathrm{ES}_{I;\alpha}$ process increases with the stability index decreasing. The proof needs to define some universal search space and the efficacy ratio of an evolutionary algorithm.

Now, the question is the following: Is it an attempt to discredit the famous *no free lunch theorem* (Wolpert and Macready 1994), which states that any two optimization algorithms are equivalent when their performance is averaged across all possible problems within a class of continuous functions? The combinatorial proof of the above theorem is based on the assumption that any optimization problem can appear with the same probability; in other words, there is a uniform probabilistic measure defined on the space of continuous functions. In our case, however, we distinguish a subclass of continuous functions contained in some finite hypercube; thus, we consider a specialized class of algorithms for some subclass of optimization problems.

5.4.1 General Searching Space

Let $\mathscr{C}[\mathscr{S}]$ represent a class of all continuous functions on $\mathscr{S} \subset \mathbb{R}^n$. Let

$$\mathscr{F}_N = \Big\{ h(\boldsymbol{x}) = \sum_{i=1}^{N} w_i \exp\big(- b_i \|\boldsymbol{x} - \boldsymbol{m}_i\|^2\big);\ w_i, b_i \in \mathbb{R}_+, \boldsymbol{m}_i \in \mathscr{S} \Big\}. \quad (5.22)$$

The well-known Stone–Weierstrass theorem (Cullen 1968) states that the set $\mathscr{F} = \bigcup_{N=1}^{\infty} \mathscr{F}_N$ is the dense set in $\mathscr{C}[\mathscr{S}]$. This means that each function $g \in \mathscr{C}[\mathscr{S}]$ can be approximated by a function $h \in \mathscr{F}$ with any accuracy, i.e.,

$$\forall \varepsilon > 0 \ \ \forall g \in \mathscr{C}[\mathscr{S}] \ \ \exists h \in \mathscr{F} \ \ \forall \boldsymbol{x} \in \mathscr{S}: \qquad \big| g(\boldsymbol{x}) - h(\boldsymbol{x}) \big| < \varepsilon. \quad (5.23)$$

The set $\mathscr{F}$ will be treated as a testing environment for the $(1+1)\mathrm{ES}_{I;\alpha}$ strategy. The Stone–Weierstrass theorem allows us to assume that conclusions obtained during the research into evolutionary strategy effectiveness over the set $\mathscr{F}$ can be, without loss of generality, related to whole class $\mathscr{C}[\mathscr{S}]$ of continuous functions.

If the set of parameters $N \in \mathbb{N}$, $w_i, b_i \in \mathbb{R}$, $i = 1, \ldots, N$, $\boldsymbol{m}_i \in \mathscr{S} \subset \mathbb{R}^n$, $i = 1, \ldots, N$, is treated as random variables, then we have stochastic parametrization of the set $\mathscr{F}$. Probabilistic distributions for individual parameters are chosen as in the works of Prętki (2008) or Prętki and Obuchowicz (2008):

- $N \in \mathbb{N}$ has Poisson's distributions with the expectation value λ_N and the probability function
$$P(N = k; \lambda_N) = \frac{\lambda_N^k \exp(-\lambda_N)}{k!}; \quad (5.24)$$
- $\boldsymbol{m}_i \in \mathscr{S}$, $i = 1, \ldots, N$, have a uniform distribution over the space $\mathscr{S}$ with the probability density function in the form
$$p(\boldsymbol{m}_i) = \frac{1}{|\mathscr{S}|}; \quad (5.25)$$
- $w_i \in \mathbb{R}$, $i = 1, \ldots, N$, and $b_i \in \mathbb{R}$ $i = 1, \ldots, N$, have an exponential distribution with the common expectation value $E[w_i] = \lambda_w$, $E[b_i] = \lambda_b$, $i = 1, \ldots, N$, and the probability density function
$$p(t; \lambda) = \begin{cases} \frac{1}{\lambda}\exp(-\frac{1}{\lambda}t), & t \geq 0, \\ 0, & t < 0. \end{cases} \quad (5.26)$$

Indicating $\Omega_k \ni \boldsymbol{\theta}_k = [N, w_1, \ldots, w_N, b_1, \ldots, b_N, \boldsymbol{m}_1, \ldots, \boldsymbol{m}_N]$, where $\Omega_k = \mathbb{N} \times \mathbb{R}^k \times \mathbb{R}^k \times \mathbb{R}^{k\times n}$, we obtain the parametric space $\Omega = \bigcup_{k=1}^{\infty} \Omega_k$ of the set of continuous functions of the form (5.22). The probabilistic measure P over the space Ω is defined by (5.24)–(5.26) and characterized by the following conditional probability distributions:

$$P(w_i \in I_w | \boldsymbol{\theta}(1) = k) = \begin{cases} \int_A p(t; \lambda) dw_i & \text{for } i \leq k, \\ 0 & \text{for } i > k, \end{cases} \tag{5.27}$$

$$P(b_i \in I_b | \boldsymbol{\theta}(1) = k) = \begin{cases} \int_A p(t; \lambda) db_i & \text{for } i \leq k, \\ 0 & \text{for } i > k, \end{cases} \tag{5.28}$$

$$P(\boldsymbol{m}_i \in I_{\boldsymbol{m}} | \boldsymbol{\theta}(1) = k) = \begin{cases} |I_{\boldsymbol{m}}| & \text{for } i \leq k, \\ 0 & \text{for } i > k, \end{cases} \tag{5.29}$$

where $I_w, I_b \subset \mathbb{R}$, $I_{\boldsymbol{m}} \subset \mathscr{S}$, while $p(\cdot; \lambda)$ is in the form of (5.26). The choice of the above probability distributions for parameters of the set $\mathscr{F}$ is not the only one possible. Most important is the fact that supports of these distributions cover the whole domain of parameters from the set $\mathscr{F}$. Due to the above fact, the following condition is met:

$$\forall g \in \mathscr{F}, \forall \delta > 0, \forall \boldsymbol{x} \in \mathscr{S} \quad P(\boldsymbol{\theta} \in \Omega : |g(\boldsymbol{x}) - h(\boldsymbol{x}; \boldsymbol{\theta})| < \delta) > 0, \tag{5.30}$$

where $h(\cdot, \boldsymbol{\theta})$ is an element of the probability space $\mathscr{F}$ corresponding to parameters $\boldsymbol{\theta}$. So, we can generate an optimization task in the testing environment approximating any continuous optimization problem with any accuracy, because the inequality (5.30) guarantees that the probability of creation of any closely approximation of any element g is higher then zero. Due to the chosen distributions (5.24)–(5.26), we can describe the set $\mathscr{F}$, which is dense in the set of all positive continuous functions, using only three parameters: λ_N, λ_w and λ_b. Values of these parameters define the probability measure over the set Ω. Thus, each triple $(\lambda_N, \lambda_w, \lambda_b)$ highlights another subclass of the class $\mathscr{C}[\mathscr{S}]$. Two sample realizations of the obtained testing environments are presented in Fig. 5.8.

5.4.2 Isotropic Stable Mutation Effectiveness in $\mathscr{F}$

Let $\boldsymbol{\theta}_N = [\boldsymbol{w}^T, \boldsymbol{b}, \boldsymbol{m}_1, \ldots, \boldsymbol{m}_N] \in \Theta_N$, where $\boldsymbol{w}^T = [w_1, \ldots, w_N] \in \mathbb{R}_+^N$, $\boldsymbol{b}^T = [b_1, \ldots, b_N] \in \mathbb{R}_+^N$, be a vector of all parameters defining the function $f \in \mathscr{F}$. Let $H_A(\boldsymbol{\xi})$ be the effectiveness of the algorithm $A(\boldsymbol{\xi})$ for an optimization problem of the function $f \in \mathscr{F}$, with $\boldsymbol{\xi}$ representing the vector of control parameters of an algorithm. The mean algorithm effectiveness over the set $\mathscr{F}$ can be obtained based on the following relation:

$$\overline{H}_A(\boldsymbol{\xi}) = \int_{\Omega} H_A(\boldsymbol{\xi}, \boldsymbol{\theta}) d\boldsymbol{\theta}. \tag{5.31}$$

It must be stressed that the above quality function has to be limited, e.g., $0 \leq H_A(\cdot) \leq 1$. It can be chosen in the form (for a maximization problem)

$$H_A(\boldsymbol{\xi}, \theta) = \frac{\max\{A(\boldsymbol{\xi}, f)\}}{f^*}, \tag{5.32}$$

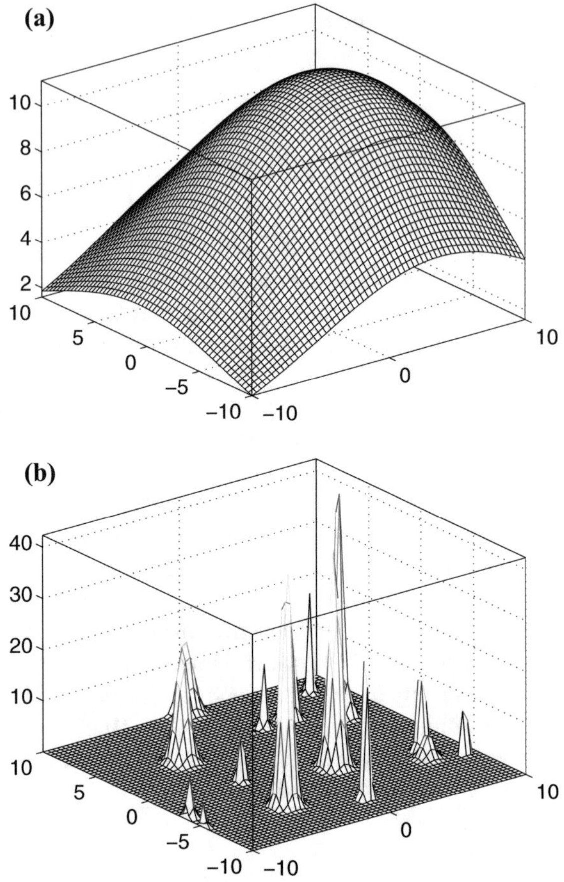

Fig. 5.8 Sample testing functions belonging to the general searching space obtained as a result of random realization of the probability measure with the following parameters: $\lambda_w = 10$, $\lambda_b = 0.01$, $\lambda_N = 0.01$ (**a**); $\lambda_w = 20$, $\lambda_b = 20$, $\lambda_N = 20$ (**b**)

where f^* is the global maximum of the objective function and $\max\{A(\boldsymbol{\xi}, f)\}$ is the best solution obtained by an algorithm A with parameters $\boldsymbol{\xi}$.

Equation (5.31) concerns deterministic algorithms, i.e., providing the same results for the same control parameters $\boldsymbol{\xi}$. If stochastic algorithms are taken into account, then the quality index H_A becomes a random variable and we should, instead of (5.31), use some expectation value:

$$E\left[H_A(\boldsymbol{\xi})\middle|\boldsymbol{\theta}\right] = \int_0^1 H_A(\boldsymbol{\xi}, \boldsymbol{\theta})\,\Pr(dH), \tag{5.33}$$

where $\Pr(\cdot)$ is the probability measure described by the $H_A(\boldsymbol{\xi}, \boldsymbol{\theta})$ values distribution for the problem from the set $\mathscr{F}$, determined by $\boldsymbol{\theta}$. Because the probability distribution of H_A is usually unknown, the integral (5.33) cannot be calculated analytically. Using the Monte-Carlo integration, the following estimator is obtained:

$$\hat{H}_A(\boldsymbol{\xi}|\boldsymbol{\theta}) = \frac{1}{N}\sum_{i=1}^{N} H_A^{(i)}(\boldsymbol{\xi}, \boldsymbol{\theta}), \tag{5.34}$$

where $H_A^{(i)}(\boldsymbol{\xi}, \boldsymbol{\theta})$ are the results (5.32) obtained by an algorithm A during its N independent processes.

Application of the Monte-Carlo method realizes the convergence of the estimator (5.34) to a real value of the integral (5.33). In practice, this convergence does not play an important role, because the estimation obtained after a limited number of random samples is always important for us. Thus, the crucial problem is to control uncertainty corresponding to the random variable (5.33). Although its distribution is usually unknown, one of the basic theorem of a statistical learning can be applied—the so-called Chernoff bound (Vidyasagar 2001), which, in our case, has the following form:

$$P^k\Big\{\boldsymbol{\theta}_N \in \Theta_N : \Big|E[H_A(\boldsymbol{\xi})] - \hat{E}[H_A(\boldsymbol{\xi})]\Big| > \varepsilon\Big\} \leq 2\exp(-2\,k\,\varepsilon^2), \tag{5.35}$$

where $\varepsilon > 0$. If the condition that the values (5.33) be included in the interval $\big[\hat{E}[H_A(\boldsymbol{\xi})] - \varepsilon, \hat{E}[H_A(\boldsymbol{\xi})] + \varepsilon\big]$ for a confidence level $1 - \delta$ is to be met, then the estimator (5.34) should be obtained based on k independent random samples $\{\sigma^{(i)}, \boldsymbol{x}_0^{(i)}, \boldsymbol{\theta}^{(i)}\}_{i=1}^{k}$, where

$$k \geq \frac{1}{2\varepsilon^2}\ln\Big(\frac{2}{\delta}\Big). \tag{5.36}$$

5.4.3 Simulation Experiments

Simulations were carried out for seven optimization algorithms: four evolutionary strategies $(1+1)\text{ES}_{I;\alpha}$ with different stable indices ($\alpha = 2, 1.5, 1, 0.5$), as well as three standard deterministic methods implemented in IMSL Fortran 90 MP Library—the quasi-Newton, Newton and Nelder–Mead methods. The experiment is performed as follows. First, optimization landscapes are created in a two-dimensional general search space determined on the set $\mathscr{S} = [-10, 10] \times [-10, 10]$ in accordance with the distributions (5.24)–(5.26). Next, each algorithm is initiated from the same chosen starting point $\boldsymbol{x}_0$. The stop criterion for the $(1+1)\text{ES}_{I;\alpha}$ strategies is defined as follows: the process has achieved a satisfactory solution or there is no progression during a hundred consecutive iterations. In the case of deterministic algorithms, stop criterions are described in detail by Bonnans et al. (2003). The entire experiment comprises $k = 73778$ independent runs of each algorithm. This value results from the inequality (5.35) for the confidence interval determined by $\varepsilon = 0.005$ and the level $1 - \theta = 0.95$. The estimate of the expectation value of the optimization effectiveness (5.34) is presented in Table 5.4. Moreover, the mean number of fitness function evaluations needed to stop the algorithms' run is shown in Table 5.5.

Table 5.4 Expectation value of the optimization effectiveness $\hat{H}_A(\boldsymbol{\xi}|\boldsymbol{\theta})$ of algorithms obtained based on the Monte-Carlo integration (5.34). $\alpha = \{2.0, 1.5, 1.0, 0.5\}$ represent the $(1+1)\mathrm{ES}_{I;\alpha}$ strategies, A_1 stands for the quasi-Newton method with the analytical form of the gradient, A_2 represents the Nelder–Mead method, A_3 is Newton's method with the Hessian analytically defined

λ_w	λ_b	λ_N	$\alpha = 2.0$	$= 1.5$	$= 1.0$	$= 0.5$	A_1	A_2	A_3
10	0.01	0.01	0.999	0.999	0.999	1.000	1.000	1.000	1.000
1	1	0.1	0.978	0.987	0.991	0.997	0.785	0.998	0.785
1	1	1	0.906	0.918	0.929	0.942	0.724	0.932	0.725
1	1	5	0.622	0.646	0.692	0.745	0.516	0.680	0.519
1	0.1	10	0.825	0.850	0.891	0.933	0.814	0.904	0.806
5	5	10	0.468	0.493	0.535	0.588	0.375	0.505	0.380
10	10	10	0.436	0.455	0.486	0.527	0.351	0.468	0.354
20	20	20	0.373	0.393	0.426	0.467	0.289	0.399	0.291

Table 5.5 Mean number of fitness function evaluations needed to achieve the results presented in Table 5.4. Results are rounded to integer numbers

λ_w	λ_b	λ_N	$\alpha = 2.0$	$\alpha = 1.5$	$\alpha = 1.0$	$\alpha = 0.5$	A_1	A_2	A_3
10	0.01	0.01	352	350	291	297	30	75	141
1	1	0.1	257	259	241	273	41	84	63
1	1	1	261	265	245	275	41	84	65
1	1	5	274	279	261	295	48	86	84
1	0.1	10	295	303	274	308	47	84	149
5	5	10	250	256	245	286	43	93	68
10	10	10	239	245	239	278	42	97	61
20	20	20	231	240	233	280	42	97	59

Tables 5.4 and 5.5 contain results of the algorithms considered for a series of eight quite different optimization landscapes. Two extreme cases: an objective function composed of one wide Gaussian peak ($\lambda_w = 10$, $\lambda_b = 0.01$, $\lambda_N = 0.01$) and an extremely multi-modal environment composed of narrow Gaussian peaks rising above an almost flat ground ($\lambda_w = 20$, $\lambda_b = 20$, $\lambda_N = 20$) are presented in Fig. 5.8. Analyzing the results from Table 5.4, it is easy to observe that the evolutionary strategies considered, despite population reduction to two elements, are significantly more effective in comparison to deterministic algorithms in the case of multi-modal environments. Deterministic algorithms using local information (first and second derivatives) localize the nearest local extremum; they cannot find the global one. Thus, they are very effective in the case of the unimodal objective function, whose results are presented in the first row in Table 5.4. Comparing the results only for evolutionary strategies, it is surprising that those with mutation based on heaviest tail distributions achieve better results for all landscapes considered, including unimodal ones, which prefer local optimization algorithms. These results are in opposition to the commonly accepted opinion that Gaussian mutation guarantees better convergence

to local extrema than, for example, Cauchy one (Rudolph 1996). Moreover, results for isotropic stable mutations are ordered as stability indices α. In other words, a lower stability index guarantees better convergence of the evolutionary strategy. The above results contradict the validity of Gaussian mutation in evolutionary algorithms.

5.5 Adaptation of Stable Mutation Parameters

Isotropic exploration distributions are one of the simplest mechanisms applied in stochastic global optimization techniques. The class of isotropic distributions is equipotent with that of positive random variables. In practice, most known random vector generators use only one, two or three parameters for unambiguous description of their distribution (Fang et al. 1990; Rudolph 1997) (e.g., the normal distribution is described only by the scale parameter σ, stable distributions additionally need the stability index α). Independently of the isotropic distribution's degrees of freedom, this model allows representing only one correlation between decision variables and the fitness function. Using an isotropic probabilistic model, the dependence between objective function values and the distance from the symmetry center can only be reflected. Thus, the exploration ability of an evolutionary algorithm with isotropic mutation in optimization landscapes with more complicated relations between variables is very limited.

One of the most important problems of $IS\alpha S(\sigma)$ distribution application in a mutation operator of evolutionary algorithms is a suitable choice of the scale parameter σ and stability index α. If no heuristic knowledge about an optimization problem is known, the task of selecting both parameters is solved by multiple tests of their various configurations, and then the choose of the best one. Such a technique is connected with high uncertainty of parameter selection, especially in the case of a low number of samples. Moreover, sometimes it cannot be performed. Also, the population in an evolutionary algorithm shifts towards areas of better fitness. During this hike, the optimization landscape changes, and so do relations between decision variables and an objective function. Therefore, a technique which has the possibility to adapt parameters of an exploration distribution during a searching process is needed.

5.5.1 *Adaptation Strategy for* $(1+1)ES_{I;\alpha}$

The stability index plays an important role in the class of distributions considered. Probability distributions with different α are characterized by quite different statistical properties. This is evidenced by theorems set forth in Chap. 3. They show that the stability index usually describes limitations of given statistical lows. Thus, adaptation procedures should be different for various values of the stability index (Rudolph 1997).

The previously mentioned 1/5 role of success (Rechenberg 1973) is based on the strong dependence between the local convergence coefficient and the so-called probability of success (Beyer and Schwefel 2002):

$$P_s = P(f(\boldsymbol{x}) > f(\boldsymbol{x} + \sigma \boldsymbol{X})), \tag{5.37}$$

where $\boldsymbol{x}_k$ is a point in the searching space, $f(\cdot)$ is an objective function and $\boldsymbol{X}$ is an isotropic interfering vector. Precise calculation of (5.37) is possible only in some special cases (Beyer and Schwefel 2002; Rudolph 1997). Analytical research of the convergence coefficient for spherical and corridor functions (Beyer 1996, 2001; Rudolph 1997) shows that, in both cases, the evolutionary strategy $(1 + 1)$ES has the quickest convergence for similar probabilities of success $P_s \approx 0.184$ (Rechenberg 1973). This result is the basis of the adaptation rule that a simple heuristic has to maintain P_s (5.37) around 1/5. This role has been obtained for relatively simple optimization landscapes and is difficult to accept for more complicated ones. Moreover, this procedure has been created for the normal distribution, which significantly differs from other stable distributions. Generally, the adaptation heuristic of 1/5-success can be described as follows:

$$\sigma_{k+1} = \begin{cases} \sigma_k / a & \text{if} \quad \hat{P}_s(\sigma_k) > p_s, \\ \sigma_k \, a & \text{if} \quad \hat{P}_s(\sigma_k) < p_s, \\ \sigma_k & \text{if} \quad \hat{P}_s(\sigma_k) = p_s, \end{cases} \tag{5.38}$$

where $\hat{P}_s(\sigma_k)$ is the probability of success estimated during algorithm processing for the scale parameter σ_k, a defines the parameter describing the speed of changes, and P_s is the optimal probability of success for the exploration distribution considered. The quantity $\hat{P}_s(\sigma_k)$ is calculated based on m observed results of succeeding mutations, i.e.,

$$\hat{P}_s(\sigma_k) = \frac{1}{m} \sum_{i=1}^{m} I(\boldsymbol{x}_{k+i} + \sigma_k \boldsymbol{Z}_i), \tag{5.39}$$

where $I(\cdot)$ is a pointer function which is equal to 1 if the solution becomes better in the ith iteration and 0 otherwise.

Another question connected with the general adaptation heuristic (5.38) is the choice of optimal method parameters in the case of application to the $IS\alpha S(\sigma)$ distributions. The answer is difficult, because optimal values of the parameters $\{a, m, p_s\}$ change for different optimization problems even in the case of the normal distribution (Beyer 2001). Therefore, in this section, we try to configure the procedure (5.38) for the $(1 + 1)\text{ES}_{I,\alpha}$ strategies separately for various most typical optimization environments in order to obtain the best possible effectiveness of extremum localization.

5.5.1.1 Simulation Experiments

Searching for the optimal configuration of the adaptation heuristic (5.38) for the strategies $(1+1)\mathrm{ES}_{I,\alpha}$ will be based on seven three-dimensional $n=3$ testing functions described in Appendix B, which can be grouped into four classes:

- well-conditioned local optimization problem: spherical function $f_{sph}(\boldsymbol{x})$ (B.1),
- badly conditioned local optimization problems: Rosenbrock's function $f_{GR}(\boldsymbol{x})$ (B.2) and elliptic function f_e (B.6),
- non-differentiable unimodal function f_{step} (B.7),
- problems with many local extrema: Rastringin's function f_R (B.4), Ackley's function f_A (B.3) and Griewank's function f_G (B.5).

The given set of testing functions $\Phi = \{f_{sph}, f_{GR}, f_e, f_R, f_A, f_G\}$ was used in many important publications (Beyer and Schwefel 2002; Yao and Liu 1996, 1999). The choice of functions used in the experiment is not really random. These functions illustrate many typical problems which are known from engineering practice. Moreover, it is worth noting that the set of testing functions contains problems which need quick narrowing of the exploration distribution (f_{sph}, f_{GR}), as well as those which prefer strong disturbances. Thus, such a set of testing functions has to facilitate the choice of $\{a, m, p_s\}$ which allows achieving a compromise between two contrary tasks: exploration and exploitation of the searching space. In order to verify final conclusions, two multimodal problems, for which the optimization success will be reached if only both properties are accommodated, are finally used in experiment.

The wide panoply of the stability indices $\alpha = 2.0, 1.75, 1.5, 1.25, 1.0, 0.75, 0.5$ is employed in this experiment. The space of steering parameters is limited to the discrete space $\{a, m, p_s\} \in A \times M \times P_s$, where

$$A = \{0.5, 0.6, 0.7, 0.8, 0.9, 0.95\}, \tag{5.40}$$

$$M = \{2, 4, 8, 16, 32, 64\}, \tag{5.41}$$

$$P_s = \{0.1, 0.15, 0.2, 0.25, 0.30, 0.35, 0.4\}. \tag{5.42}$$

Each combination of the above parameters and stability indices is considered during the experiment, so we have 1764 different configurations of the $(1+1)\mathrm{ES}_{I;\alpha}$ strategies. Because the result of an evolutionary process is indeed a random variable, some quality indicator should be defined over the result distribution. A two-criterion quality indicator, i.e., the median of the final value of an objective function Me[WFC] and the median of the number of generations needed to achieve the stop criterion Me[LOFC], is chosen in our experiment. However, the median is a statistic parameter which is robust to occurrence of extreme values of a statistical feature, and 100 independent evolutionary algorithm processes are used in order to calculate the medians considered. In this way, taking also into account the number of testing objective functions, the results are obtained after 1234800 independent runs of the evolutionary strategy $(1+1)\mathrm{ES}_{I;\alpha}$.

Each test is processed as follows. Each $(1+1)\mathrm{ES}_{I;\alpha}$ algorithm is initiated from the same point in the searching space with the same initial scale parameter $\sigma = 1$ and stopped after fulfilling one of the following three conditions:

1. a result of an acceptable quality (i.e., lower than 10^{-5}) is found;
2. the number of objective function evaluations exceeds the limit value $T_{\max} = 10000$;
3. the value of the scale parameter prevents effective space searching, i.e., $\sigma < 10^{-10}$.

In order to find the best parameter configuration, some order in the set $A \times P_s \times M$ is established according to the following rule: the combination $\omega^{(1)} = \{a^{(1)}, m^{(1)}, p_s^{(1)}\}$ is better than $\omega^{(2)} = \{a^{(2)}, m^{(2)}, p_s^{(2)}\}$ when

(a) the median Me[WFC] for $\omega^{(1)}$ is lower;
(b) if both medians Me[WFC] for $\omega^{(1)}$ and $\omega^{(2)}$ are equal (this is the situation when medians achieve the threshold value for both configurations), the configuration with a lower median Me[LOFC] is chosen as better.

In order to simplify the analysis, let us introduce some auxiliary function $R(\omega, \phi)$:

$$R : A \times M \times P_s \times \Phi \to [1, \ldots, 252], \tag{5.43}$$

which gives the order number of the combination $\omega = \{a, m, p_s\}$ in the series established by the results obtained for a testing function $\phi \in \Phi$.

The best parameter combinations for 3D testing functions are presented in Table 5.6.

Analyzing optimal configurations obtained during the experiment (Table 5.6), the most significant fact which can be noticed is domination of the distribution with the stable index $\alpha = 0.5$. However, this algorithm usually needs more iterations than, for example, $(1+1)\mathrm{ES}_{I;2}$, but a properly configured strategy $(1+1)\mathrm{ES}_{I;0.5}$ guarantees global extrema localization. Algorithms with higher stability indices $\alpha = 2, 1.5$ usually early converge to a local extremum (e.g., f_A). Large differentiation of optimal parameters for adaptation heuristics can prove large differentiation of topological characteristics of the testing functions applied. It is interesting that the heuristic guaranteeing rapid and frequent scale parameter modifications turns out to be best for the spherical function f_{sph} regardless of the stability index. Large values of m for optimal adaptation heuristics obtained for multimodal problems prove that the effectiveness of the evolutionary strategy for these problems is conditioned by slow modifications of σ.

Observing radically different adaptation procedure configurations, it can be supposed that possible changes of individual parameters influence an evolutionary strategy is effectiveness for different optimization tasks. The analysis of such a type of relations can be performed as follows. Parameters of an adaptation heuristic can be treated as explanatory variables and values of the function R (5.43) as the dependent variable. Table 5.7 presents linear correlation coefficients for individual stability indices.

Table 5.6 Optimal values of the parameters $\{a, m, p_s\}$ for each testing function. Space dimension $n = 3$

$\alpha = 2$					
	a	m	p_s	Me[WFC]	Me[LOFC]
f_{sph}	0.5	2	0.25	$4.5414e-006$	132
f_{GR}	0.95	64	0.15	11.841	10000
f_{step}	0.5	4	0.15	0	109
f_e	0.9	32	0.1	$8.9243e-006$	2171.5
f_R	0.9	32	0.15	$7.5737e-006$	2148
f_A	0.5	2	0.1	19.9668	144
f_G	0.5	64	0.1	0.029584	2944
$\alpha = 1.5$					
	a	m	p_s	Me[WFC]	Me[LOFC]
f_{sph}	0.5	2	0.1	$6.3974e-006$	135.5
f_{GR}	0.95	64	0.1	13.0852	10000
f_{step}	0.6	4	0.1	0	116.5
f_e	0.95	16	0.2	$9.8451e-006$	2657
f_R	0.9	64	0.15	$9.4678e-006$	3595.5
f_A	0.5	2	0.15	19.9668	140
f_G	0.6	64	0.1	0.029584	3840
$\alpha = 1$					
	a	m	p_s	Me[WFC]	Me[LOFC]
f_{sph}	0.5	2	0.2	$5.3288e-006$	143.5
f_{GR}	0.95	64	0.15	8.9948	10000
f_{step}	0.6	4	0.15	0	118
f_e	0.9	64	0.4	$5.9589e-006$	3178
f_R	0.9	64	0.1	$8.8239e-006$	3289.5
f_A	0.9	64	0.3	$9.0356e-006$	5921.5
f_G	0.5	64	0.15	0.029584	2816
$\alpha = 0.5$					
	a	m	p_s	Me[WFC]	Me[LOFC]
f_{sph}	0.5	2	0.1	$5.5879e-006$	184.5
f_{GR}	0.95	64	0.3	$9.374e-006$	7131
f_{step}	0.8	4	0.15	0	132.5
f_e	0.95	32	0.1	$8.245e-006$	3684
f_R	0.9	64	0.25	$6.7042e-006$	2728.5
f_A	0.5	32	0.1	$8.7294e-006$	860.5
f_G	0.95	64	0.15	0.0098647	10000

Table 5.7 Linear correlation coefficients between the adaptation procedure parameters $\{a, m, p_s\}$ and a position in ascending order ranking of configuration sets—the searching space dimension $n = 3$

Objective function f_{sph}

Zmienne	$\alpha = 2$	$= 1.75$	$= 1.5$	$= 1.25$	$= 1$	$= 0.75$	$= 0.5$
a	0.58	0.57	0.57	0.56	0.56	0.55	0.53
m	0.72	0.72	0.72	0.72	0.72	0.72	0.71
p_s	−0.07	−0.07	−0.07	−0.08	−0.10	−0.12	−0.17

Objective function f_{GR}

Zmienne	$\alpha = 2$	$= 1.75$	$= 1.5$	$= 1.25$	$= 1$	$= 0.75$	$= 0.5$
a	−0.37	−0.34	−0.36	−0.47	−0.23	−0.26	−0.34
m	−0.43	−0.47	−0.40	−0.36	−0.45	−0.35	−0.37
p_s	0.06	0.05	0.03	0.06	0.15	0.08	0.18

Objective function f_{step}

Zmienne	$\alpha = 2$	$= 1.75$	$= 1.5$	$= 1.25$	$= 1$	$= 0.75$	$= 0.5$
a	−0.00	−0.03	−0.02	0.01	−0.03	−0.08	−0.21
m	0.20	−0.07	−0.12	−0.09	−0.10	−0.10	−0.26
p_s	0.57	0.63	0.60	0.62	0.62	0.62	0.53

Objective function f_e

Zmienne	$\alpha = 2$	$= 1.75$	$= 1.5$	$= 1.25$	$= 1$	$= 0.75$	$= 0.5$
a	−0.38	−0.36	−0.36	−0.37	−0.33	−0.33	−0.29
m	−0.54	−0.55	−0.53	−0.51	−0.50	−0.46	−0.40
p_s	0.47	0.48	0.48	0.50	0.52	0.56	0.61

Objective function f_R

Zmienne	$\alpha = 2$	$= 1.75$	$= 1.5$	$= 1.25$	$= 1$	$= 0.75$	$= 0.5$
a	−0.41	−0.43	−0.42	−0.45	−0.47	−0.45	−0.43
m	−0.73	−0.71	−0.73	−0.72	−0.73	−0.74	−0.76
p_s	0.23	0.23	0.23	0.20	0.19	0.17	0.16

Objective function f_A

Zmienne	$\alpha = 2$	$= 1.75$	$= 1.5$	$= 1.25$	$= 1$	$= 0.75$	$= 0.5$
a	0.98	0.98	0.96	0.87	0.56	0.08	−0.14
m	0.15	0.15	0.14	−0.01	−0.29	−0.62	−0.61
p_s	0.03	0.03	0.01	0.04	0.07	0.05	0.11

Objective function f_G

Zmienne	$\alpha = 2$	$= 1.75$	$= 1.5$	$= 1.25$	$= 1$	$= 0.75$	$= 0.5$
a	−0.05	−0.22	−0.17	−0.07	0.07	−0.08	−0.38
m	−0.70	−0.61	−0.48	−0.34	−0.21	−0.36	−0.58
p_s	0.24	0.34	0.52	0.63	0.76	0.70	0.47

Two types of dependencies can be distinguished in the correlations presented in Table 5.7. One is observed for the function f_{sph} and consists of positive values of coefficients for parameters a, m. It means that the mean effectiveness of the adaptation heuristic decreases when a and m increase. Quite a different situation can be observed for other test problems. The same parameters are negatively correlated for algorithms which end the optimization process with success. Because lower values of a, m manifest themselves by the strongest and more frequent changes of the scale parameter σ, the results shown in Table 5.7 explain values of optimal configuration sets (Table 5.6).

The results presented in Table 5.6 can be the basis of a general conclusion: the evolutionary startegy $(1+1)\text{ES}_{I;0.5}$ with the adaptation heuristic with parameters $\{a \approx 0.9, m \approx 64, p_s \approx 0.15\}$ is most effective for multi-dimensional problems. For well-conditioned problems, evolutionary strategies of any stability index but with $\{a \approx 0.6, m \approx 4, p_s \approx 0.15\}$ are recommended. The observed rules are not helpful if we do not have any *apriori* information about the global optimization problem solved. The question is: Does a universal adaptation procedure configuration, which can be recommended in the general case without any additional assumptions, exist? This question is also about looking for a compromise between these two extremely different adaptation heuristics. The compromise can be obtained by the choice of a configuration set ω^*, which possesses the highest mean effectiveness over the whole testing set Φ. In the case considered, the following relation is used in order to obtain universal configurations:

$$\omega^* = \arg\min_{\omega} \sum_{\phi \in \Phi} R(\omega, \phi). \tag{5.44}$$

Thus obtained parameters of the adaptation heuristic with related results are presented in Table 5.8.

Analysing the results presented in Table 5.8, it is easy to notice that strategies with the stability index $\alpha = 0.5$ dominate. Departing from optimal configurations (Table 5.6) produces worse results, but the observed effectiveness decrease is weakest in the case of $(1+1)\text{ES}_{I;0.5}$. Another rule can also be observed: values of a and m in universal sets increase with the stability index α decreasing, and the scale parameter changes should be much violent for lower values of α.

It is worth noting that, in most cases, the optimal configurations contain parameters $\{a, m, p_s\}$ which have limit values from the assumed ranges. So, we must presume that there exist better configurations, which are out of discrete sets of parameter values considered in the presented experiment. Another problem is the generalization of our results to searching spaces of higher dimensions. Taking into account non-linear characteristics of isotropic stable distributions (Fig. 5.2), this problem seems to be rather difficult. Perhaps, the general hypothesis about slower narrowing down of heavy-tail distributions turns out to be true, but parameters of optimal configurations probably do not depend linearly on n. These problems should be solved in the future.

Table 5.8 Adaptation heuristic configurations which guarantee the best mean optimization effectiveness. Search space dimension $n = 3$

$\alpha = 2$

	a	m	p_s	Me[WFC]	Me[LOFC]
f_{sph}	0.5	4	0.1	$4.5329e-006$	199.5
f_{GR}	0.5	4	0.1	15.7917	10000
f_{step}	0.5	4	0.1	0	116
f_e	0.5	4	0.1	0.070833	10000
f_R	0.5	4	0.1	16.4167	380
f_A	0.5	4	0.1	19.9668	264
f_G	0.5	4	0.1	0.2663	452

$\alpha = 1.5$

	a	m	p_s	Me[WFC]	Me[LOFC]
f_{sph}	0.6	4	0.15	$4.419e-006$	231
f_{GR}	0.6	4	0.15	17.0921	10000
f_{step}	0.6	4	0.15	0	124
f_e	0.6	4	0.15	0.061305	10000
f_R	0.6	4	0.15	7.9597	468
f_A	0.6	4	0.15	19.9668	308
f_G	0.6	4	0.15	0.2207	536

$\alpha = 1.0$

	a	m	p_s	Me[WFC]	Me[LOFC]
f_{sph}	0.5	8	0.1	$4.3124e-006$	333.5
f_{GR}	0.5	8	0.1	16.7882	10000
f_{step}	0.5	8	0.1	0	155.5
f_e	0.5	8	0.1	$1.4115e-005$	10000
f_R	0.5	8	0.1	1.9899	752
f_A	0.5	8	0.1	19.9668	512
f_G	0.5	8	0.1	0.12202	912

$\alpha = 0.5$

	a	m	p_s	Me[WFC]	Me[LOFC]
f_{sph}	0.7	16	0.1	$3.9677e-006$	419
f_{GR}	0.7	16	0.1	10.5552	10000
f_{step}	0.7	16	0.1	0	157.5
f_e	0.7	16	0.1	$9.761e-006$	8179.5
f_R	0.7	16	0.1	1.4924	1344
f_A	0.7	16	0.1	$9.1996e-006$	995
f_G	0.7	16	0.1	0.10726	1584

5.6 Summary

An undoubted advantage of mutation based on the $IS\alpha S(\sigma)$ distribution, in comparison with that based on the $NS\alpha S(\boldsymbol{\sigma})$ distribution, is its independence of reference frame selection in the searching space. Unfortunately, the dead surrounding effect is valid. However, it can be less cumbersome for heavy-tail distributions, because, unlike in the case of non-isotropic stable mutation, the mode of the distribution of $\|\boldsymbol{X}\|$, $\boldsymbol{X} \sim IS\alpha S(\sigma)$, decreases with the stability index decreasing. It can be suspected that algorithms with soft selection and isotropic stable mutation with low values of α will localize an extremum more precisely. Taking into account that, in this case, macro-mutations are more probable, there is a chance that exploration and exploitation abilities can be balanced.

The analysis of local convergence of the evolutionary strategy $(1+1)\mathrm{ES}_{I;\alpha}$ proves the advantage of the strategy with Gaussian mutation. Strategies with mutations with low stability indices are more precise in a close neighborhood of an extremum only. This fact is connected with the above-mentioned relation between the dead surrounding and α. However, the effectiveness of an evolutionary strategy with mutations with high values of α is more sensitive to precise selection of optimal control parameters of the mutation process, especially concerning the scale parameter. Analysis of robustness of the evolutionary strategy with isotropic stable mutation to parameter selection, realized in the general searching space, confirms the superiority of algorithms with mutation of a low stability index over mutation based on the normal distribution.

The adaptation strategy of the scale parameter of isotropic stable mutation is the subject of the last subsection. The presented results draw one's attention to the choice of a suitable distribution of the $IS\alpha S(\sigma)$ class and a suitable adaptation strategy according to the type of the optimized objective function.

References

Auger, A., & Hansen, N. (2006). Reconsidering the progress rate theory for evolution strategies in finite dimensions. *8th Annual Conference Genetic and Evolutionary Computation GECCO'06* (pp. 445–452). New York: ACM.

Beyer, H. G. (1996). Toward a theory of evolution strategies: Self-adaptation. *Evolutionary Computation, 3*(3), 311–347.

Beyer, H. G. (2001). *The theory of evolution strategies*. Heidelberg: Springer.

Beyer, H. G., & Schwefel, H. P. (2002). Evolution strategies–a comprehensive introduction. *Natural Computing, 1*(1), 3–52.

Bienvenue, A., & Francois, O. (2003). Global convergence for evolution strategies in spherical problems: Some simple proofs and difficulties. *Theoretical Computation Science, 306*(1–3), 269–289.

Bonnans, J. F., Gilbert, J Ch., Lemaréchal, C., & Sagastizábal, C. (2003). *Numerical optimization*. Heidelberg: Springer.

Cullen, H. F. (1968). *Introduction to general topology*. Boston: Heath.

Fang, K.-T., Kotz, S., & Ng, K. W. (1990). *Symmetric multivariate and related distributions*. London: Chapman and Hall.

Galar, R. (1990). *Soft selection in random global adaptation in* R^n*: A biocybernetic model of development*. Wrocław (in Polish): Technical University of Wrocław Press.

Hansen, N., Gemperle, F., Auger, A., & Koumoutsakos, P. (2006). When do heavy-tail distributions help? In T. Ph. Runarsson, H. -G. Beyer, E. Burke, J. J. Merelo-Guervós, L. D. Whitley, & X. Yao (Eds.), *Problem solving from nature (PPSN) IX* (Vol. 4193, pp. 62–71). Lecture notes in computer science. Berlin: Springer.

Karcz-Dulęba, I. (2004). Time to convergence of evolution in the space of population states. *International Journal Applied Mathematics and Computer Science*, *14*(3), 279–287.

Prętki, P. (2008). α*-Stable distributions in evolutionary algorithms for global parametric optimization*. Ph.D. thesis, University of Zielona Góra (in Polish).

Prętki P., & Obuchowicz A. (2008). Robustness of isotropic stable mutations in general search space. In L. Rutkowski, R. Scherer, R. Tadeusiewicz, L. A. Zadeh, & J. M. Zurada (Eds.), *Artificial intelligence and soft computing* (Vol. 5097, pp. 460–468). Lecture notes on artificial intelligence. Berlin: Springer.

Rechenberg, I. (1973). *Evolutionsstrategie: Optimierung technischer Systeme nach Prinzipien der biologischen Evolution*. Stuttgard: Frommann-Holzburg Verlag.

Rudolph, G. (1996). Convergence of evolutionary algorithms in general search space. In *International Conference on Evolutionary Computation 1996, Nagoya, Japan* (pp. 50–56).

Rudolph, G. (1997). Local convergence rates of simple evolutionary algorithms with Cauchy mutations. *IEEE Transactions on Evolutionary Computation*, *1*(4), 249–258.

Vidyasagar, M. (2001). Randomized algorithms for robust controller synthesis using statistical learning theory. *Automatica*, *37*(10), 1515–1528.

Wolpert, D., and Macready, W. (1994). *The mathematics of search*. Technical report No. SFI-TR-95-02-010, Santa Fe Institute, Santa Fe.

Yao, X., & Liu, Y. (1997). Fast evolutionary strategies. *Control Cybernetics*, *26*(3), 467–496.

Yao, X., & Liu, Y. (1999). Evolutionary programming made faster. *IEEE Transactions on Evolutionary Computation*, *3*(2), 82–102.

Chapter 6
Stable Mutation with the Discrete Spectral Measure

In accordance with Theorem 3.13, a random stable vector spectral representation is the pair $(\Gamma_s, \boldsymbol{\mu}_0)$, where Γ_s is the so-called spectral measure of the α-stable vector X and $\boldsymbol{\mu}_0$ is a localization vector. Therefore, the application of multidimensional stable distributions to global optimization tasks is limited to two simple cases. One is (described in Chap. 4) the non-isotropic stable distribution $NS\alpha S(\boldsymbol{\sigma})$, for which the spectral measure is generated by points lying on the axis of the Cartesian reference frame. The other case is (described in Chap. 5) the isotropic stable distribution $IS\alpha S(\sigma)$, whose spectral measure is uniformly distributed on the surface of the unit sphere. If we limit ourselves only to these two cases, it may happen that many properties of stable distributions, which can be valuable in the optimization contexts, will not be discovered and used. In order to obtain the possibility of modeling the complicated dependence between decision variables, we consider the possibility of wide stable vector class generation using the *discrete spectral measure* (DSM).

6.1 Stable Distributions with the DSM

The discrete spectral measure Γ_s can be defined using the Dirac delta in the following way:

$$\Gamma_s(\cdot; \boldsymbol{\xi}, \boldsymbol{\gamma}) = \sum_{i=1}^{n_s} \gamma_i \delta_{s_i}(\cdot), \tag{6.1}$$

where $\boldsymbol{\xi} = \{\boldsymbol{s}_i \mid i = 1, 2, \ldots, n_s\}$, $\boldsymbol{s}_i \in S_n$, is the set of distribution base points distributed on the surface of the n-dimensional unit sphere and $\boldsymbol{\gamma} = [\gamma_i \mid i = 1, 2, \ldots, n_s] \in \mathbb{R}_+^{n_s}$ is the corresponding weight vector. In this way, the probabilistic measure of each subset $A \subset S_n$ is defined as follows:

A. Obuchowicz, *Stable Mutations for Evolutionary Algorithms*,
Studies in Computational Intelligence 797,
https://doi.org/10.1007/978-3-030-01548-0_6

$$\Gamma_s(A) = \sum_{i=1}^{n_s} \gamma_i I_A(\boldsymbol{s}_i), \tag{6.2}$$

where $I_A(\cdot)$ is the indicator function of the subset A. Based on Theorem 3.13, the characteristic function of the random stable vector described by the DSM (6.1) has the form (Nolan et al. 2001)

$$\varphi(\boldsymbol{k}) = \exp\left(-\sum_{i=1}^{n_s} \gamma_i |\boldsymbol{k}^T \boldsymbol{s}_i|^\alpha \left(1 - i\,\mathrm{sgn}(\boldsymbol{k}^T \boldsymbol{s}_i) \tan\left(\frac{\pi\alpha}{2}\right)\right) + i\boldsymbol{k}^T \boldsymbol{\mu}\right) \tag{6.3}$$

for $\alpha \neq 1$ and

$$\varphi(\boldsymbol{k}) = \exp\left(-\sum_{i=1}^{n_s} \gamma_i |\boldsymbol{k}^T \boldsymbol{s}_i| \left(1 - i\frac{2}{\pi}\mathrm{sgn}(\boldsymbol{k}^T \boldsymbol{s}_i) \ln|\boldsymbol{k}^T \boldsymbol{s}_i|\right) + i\boldsymbol{k}^T \boldsymbol{\mu}\right) \tag{6.4}$$

for $\alpha = 1$.

The definition of the DSM allows generating multidimensional stable distributions in a very simple way. One can notice that the application of the DSM does not introduce any limits on general properties of multidimensional random stable vectors. This results from the following theorem, whose proof can be found in a publication by Byczkowski et al. (1993).

Theorem 6.1 *Let $p(\boldsymbol{x})$ be a probability density function of the stable distribution described by the characteristic function defined in Theorem 3.13, and $p^*(\boldsymbol{x})$ be a probability density function of the random vector described by the characteristic function* (6.3) *and* (6.4). *Thus,*

$$\forall\varepsilon > 0 \;\; \exists n_s \in \mathbb{N} \;\; \exists \boldsymbol{\xi}, \boldsymbol{\gamma} \;\; \forall \boldsymbol{x} \in \mathbb{R}^n : \quad \sup_{\boldsymbol{x}\in\mathbb{R}^n} |p(\boldsymbol{x}) - p^*(\boldsymbol{x})| < \varepsilon. \tag{6.5}$$

In other words, any stable distribution can be approximated by some distribution based on the DSM with any accuracy.

An important problem is to develop an effective (in the sense of computational complexity) pseudo-random generator of stable vectors. The following property of the stochastic decomposition is useful for this task (Modarres and Nolan 2002):

$$\boldsymbol{X}_{\boldsymbol{\xi}}^{\boldsymbol{\gamma}} \stackrel{d}{=} \begin{cases} \sum_{i=1}^{n_s} \gamma_i^{1/\alpha} Z_i \boldsymbol{s}_i & \text{for } \alpha \neq 1, \\ \sum_{i=1}^{n_s} \gamma_i^{1/\alpha} \left(Z_i - \frac{2}{\pi}\ln(\gamma_i)\right) \boldsymbol{s}_i & \text{for } \alpha = 1, \end{cases} \tag{6.6}$$

where Z_i are i.i.d. random stable variables of the $S_\alpha(1, 1, 0)$ distribution, for which an effective generator is described by Theorem 3.10, $\boldsymbol{\gamma}$ is a weight vector, $\boldsymbol{\xi} = \{\boldsymbol{s}_i \mid i = 1, 2, \ldots, n_s\}$, $\boldsymbol{s}_i \in S_n$, is some set of basis vectors (the spanning set).

If the DSM with the spanning set $\boldsymbol{\xi}$ composed of unit vectors in a Cartesian coordinate system and the weight vector $\boldsymbol{\gamma}$ composed of all ones are used to generate a random vector, then this vector is of the $NS\alpha S(\sigma)$ (3.47) distribution, i.e., each component of the random vector is of the symmetric stable distribution $S\alpha S(\sigma)$. The generator described by (6.6) and using the discrete spectral measure broadens each component of the random vector into two additional degrees of freedom described by parameters β and μ, i.e., each component of the vector $\boldsymbol{X}$ is of the $S_\alpha(\sigma, \beta, \mu)$ distribution. The above thesis can be concluded from the following theorem, proved by Samorodnitski and Taqqu (1994).

Theorem 6.2 *The spectral measure of the stable vector* $\boldsymbol{X} = [X_i \mid i = 1, 2, \ldots, n]^T$ *is described by the finite number of spanning vectors* $\boldsymbol{\xi} = \{\boldsymbol{s}_i \mid i = 1, 2, \ldots, n_s\}$, $\boldsymbol{s}_i \in S_n$, *if and only if the vector* $\boldsymbol{X}$ *can represented by the linear combination of independent stable random variables, i.e.,*

$$\boldsymbol{X} = \boldsymbol{A}\boldsymbol{Z}, \tag{6.7}$$

where $\boldsymbol{A} \in \mathbb{R}^{n \times n_s}$, $\boldsymbol{Z} = [Z_k \sim S_\alpha(\sigma, \beta, \mu) \mid k = 1, 2, \ldots, n_s]^T$.

In light of the above, the DSM can be used to describe the statistical dependence between decision variables, whose detection and inclusion into the mutation operator distinctly accelerate an optimization process.

In terms of optimization procedures, one of the most essential properties of distributions based on the DSM is described by Theorem 3.16 and concerns the asymptotic form of multidimensional distribution tails. In the case of a spectral measure representation in the form of a discrete set of points, it means that the whole probability mass, which is significantly away from the base point, will be focused on directions parallel to vectors from the set $\boldsymbol{\xi}$. Thus, macro-mutations can occur only along vectors spanning the DSM. This effect is illustrated in Fig. 6.1.

The influence of the DSM on the probability distribution form is clearly illustrated by the multidimensional normal distribution $\mathbf{N}(\boldsymbol{\mu}, \boldsymbol{\Sigma})$ with the characteristic function

$$\varphi(\boldsymbol{k}) = \exp\left(-\frac{1}{2}\boldsymbol{k}^T \boldsymbol{\Sigma} \boldsymbol{k} + i\boldsymbol{\mu}^T \boldsymbol{k}\right). \tag{6.8}$$

Let us apply the singular-value decomposition (SVD) to the covariance matrix, i.e., $\boldsymbol{\Sigma} = \boldsymbol{U}\,\boldsymbol{\Lambda}\,\boldsymbol{U}$. In order to transform the function (6.8) to the form appropriate for the DSM (Theorem 3.13), it is sufficient to adapt

$$\boldsymbol{U} = [\boldsymbol{s}_1, \boldsymbol{s}_2, \ldots, \boldsymbol{s}_{n_s}], \quad \Lambda = \begin{bmatrix} 2\gamma_1 & 0 & \cdots & 0 \\ 0 & 2\gamma_2 & 0 & 0 \\ \vdots & \vdots & \vdots & \vdots \\ 0 & 0 & \cdots & 2\gamma_{n_s} \end{bmatrix}. \tag{6.9}$$

This means that eigenvectors of the covariance matrix of the normal distribution are simultaneously vectors spanning the DSM, while eigenvalues are weights of the DSM.

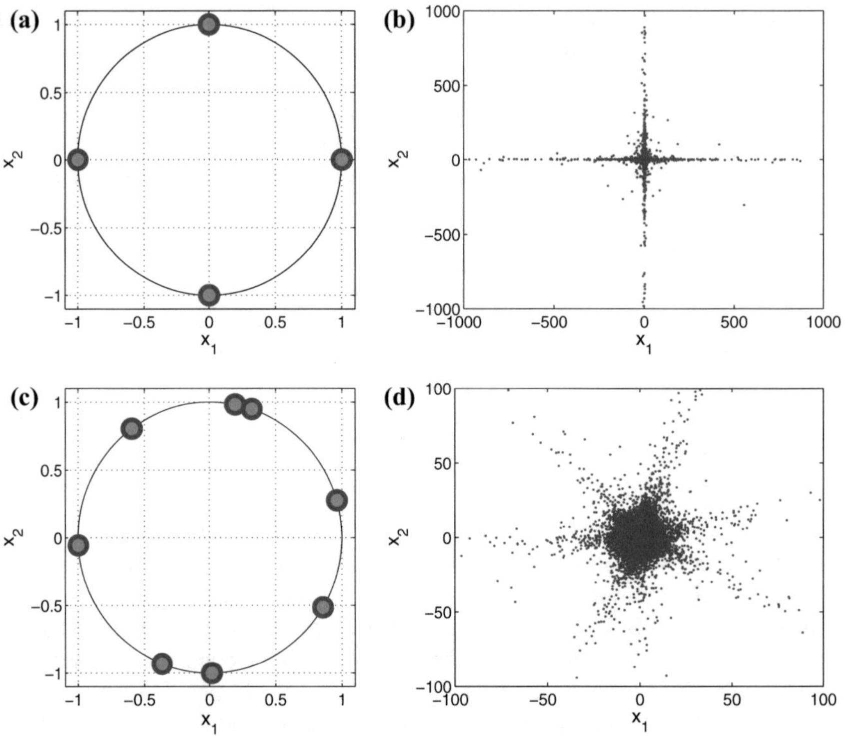

Fig. 6.1 Distribution of vectors spanning the discrete spectral measure and the corresponding random realizations: $\alpha = 0.75$ (**a**), (**b**), $\alpha = 1.5$ (**c**), (**d**)

Summarizing comments on multidimensional stable distributions, one should wonder what benefits evolutionary algorithms might gain from adopting them. In order to properly address this problem, two aspects should first be taken into account: macro-mutations, which allow crossing evolutionary saddles significantly easier, and the possibility of modelling complex stochastic dependencies, which guarantees higher local optimization effectiveness. Benefits of application of distributions described by the DSM are illustrated by the numerical experiment presented below.

6.2 Selection of the Optimal Stable Distribution

A two-dimensional Rosenbrock function (B.2)

$$f_{GR2}(\boldsymbol{x}) = (1 - x_1)^2 + 100(x_2 - x_1^2)^2 \tag{6.10}$$

is used in the numerical experiment (Obuchowicz and Prętki 2010). Let us assume that the point $\boldsymbol{x}_k = [0, -2]^T$ $(f_{GR2}(\boldsymbol{x}_k) = 401)$ is some estimation of the optimal point obtained during some previous findings. The goal is correction of the recent solution by the its random disturbance by adding a random vector $\boldsymbol{X}_{\xi}^{\gamma}$. The aim of the calculations presented below is the choice of a suitable stable model from the class of multidimensional distributions based on the DSM. First, rival probabilistic models are restricted to the set of four stable distributions $\Omega = \{\boldsymbol{X}_{\xi}^{\gamma}(\alpha)|\alpha = 0.5, 1.0, 1.5, 2.0\}$. Each random vector $\boldsymbol{X}_{\xi}^{\gamma}(\alpha)$ is described by the DSM based on 16 uniformly distributed points:

$$\xi = \left\{ \begin{bmatrix} 1 \\ 0 \end{bmatrix}, \begin{bmatrix} 0.92 \\ 0.38 \end{bmatrix}, \begin{bmatrix} 0.7 \\ 0.7 \end{bmatrix}, \begin{bmatrix} 0.38 \\ 0.92 \end{bmatrix}, \begin{bmatrix} 0 \\ 1 \end{bmatrix}, \begin{bmatrix} -0.38 \\ 0.92 \end{bmatrix}, \begin{bmatrix} -0.7 \\ 0.7 \end{bmatrix}, \begin{bmatrix} -0.92 \\ 0.38 \end{bmatrix}, \cdots \right.$$

$$\left. \cdots, \begin{bmatrix} -1 \\ 0 \end{bmatrix}, \begin{bmatrix} -0.92 \\ -0.38 \end{bmatrix}, \begin{bmatrix} -0.92 \\ 0.38 \end{bmatrix}, \begin{bmatrix} -1 \\ 0 \end{bmatrix}, \begin{bmatrix} -0.92 \\ -0.38 \end{bmatrix} \right\}.$$

The criterion of selection of the best model from the set Ω is defined as follows:

$$\gamma^* = \arg \min_{\gamma \in \mathbb{R}_+^{16}} C(\gamma), \tag{6.11}$$

where

$$C(\gamma) = E\left[\min\left\{\frac{\varphi(\boldsymbol{x}_k + \boldsymbol{X}_{\xi}^{\gamma}(\alpha))}{\varphi(\boldsymbol{x}_k)}, 1\right\}\right]. \tag{6.12}$$

Unfortunately, the function (6.12) does not have the analytical form, thus the problem (6.11) cannot be solved using standard optimization methods. One of possible ways to solve it is application of the Monte-Carlo method (Kemp 2003; MacKey 1998). The law of large numbers (Durrett 1995) allows estimating the expectation value (6.12) using the following estimator:

$$\hat{C}(\gamma) = \frac{1}{N}\sum_{l=1}^{N} \min\left\{\frac{\varphi(\boldsymbol{x}_k + \boldsymbol{X}_{i,\xi}^{\gamma}(\alpha))}{\varphi(\boldsymbol{x}_k)}, 1\right\}, \tag{6.13}$$

where $\{\boldsymbol{X}_{i,\xi}^{\gamma}(\alpha)) \mid i = 1, 2, \ldots, N\}$ is the series of independent realizations of the random vector of the α-stable distribution. Using the estimator (6.13), the problem (6.11) can be rewritten to the form

$$\gamma^* = \arg \min_{\gamma \in \mathbb{R}^{16}} \hat{C}(\gamma). \tag{6.14}$$

A suitable choice of the number of samples N is very important for such transformations. The objective function has stochastic properties and the number N allows us to control the expectation value (6.13). It would be advantageous to choose a large N, but such a choice is connected with a very large computational effort of the method. In order to obtain a rational choice of N and proper control of the estimator

Table 6.1 Objective function values and the corresponding probabilities of success obtained for pseudo-optimal stable distributions

α	2.0	1.5	1.0	0.5
$C(\gamma^*)$	0.6683	0.4978	0.6522	0.5385
P_s	0.4852	0.6445	0.5145	0.6638

(6.13), one of the theorems of the theory of probability, the so-called Chernoff bound (Vidyasagar 2001), can be applied.

Because of the stochastic character of the objective function and the necessity to achieve some compromise between the estimator quality and computational effort, standard optimization techniques are useless. Therefore, the simultaneous perturbation stochastic approximation (SPSA) algorithm (Spall 1998, 2003) is chosen as the optimization method. It is dedicated to multidimensional stochastic function optimization. The results are presented in Table 6.1. The set of optimal weights, focus points and the corresponding probability density functions are illustrated in Fig. 6.2.

The obtained results show the advantage of distributions with low values of the stability index α. It is surprising that they are much more effective than the normal distribution ($\alpha = 2$) in the case of the unimodal Rosenbrock function despite their frequent macro-mutations. The main cause of this is the spherical symmetry of the normal distribution. In this case, the probability of success cannot exceed 0.5 for most objective functions. In distributions without the spherical symmetry $\alpha < 2$, it is possible to channel exploration distributions into the area of, on average, fewer values of an objective function. This channeling is best shown if one compares the probabilities of success presented in Table 6.1. It can be clearly seen that, in the case of optimal distributions, the creation probability of a solution better than $\boldsymbol{x}_k = [0, -2]^T$ is higher than 0.60.

6.3 DSM Adaptation

The effectiveness of an optimization procedure used to solve global optimization problems significantly increases when there is a possibility to adapt its configuration parameters. This fact is particularly important in the case of mutation parameter adaptation in evolutionary algorithms. Application of static exploration distributions is limited: either sampling is too dense and the procedure is trapped around some local extremum, or sampling is too wide and the procedure has a problem with convergence. Parameter adaptation allows compromising the two effects.

A parameter adaptation method for algorithms with mutations based on multidimensional stable distributions described by the DSM is the subject of this section. The idea of the proposed technique is taken from *estimation of distribution algorithms* (EDA) (Larranaga and Lozano 2001). It uses information from the population of solutions which are an alternative for the chosen probabilistic model. First, many alternative solutions are generated around a given base solution using a mutation

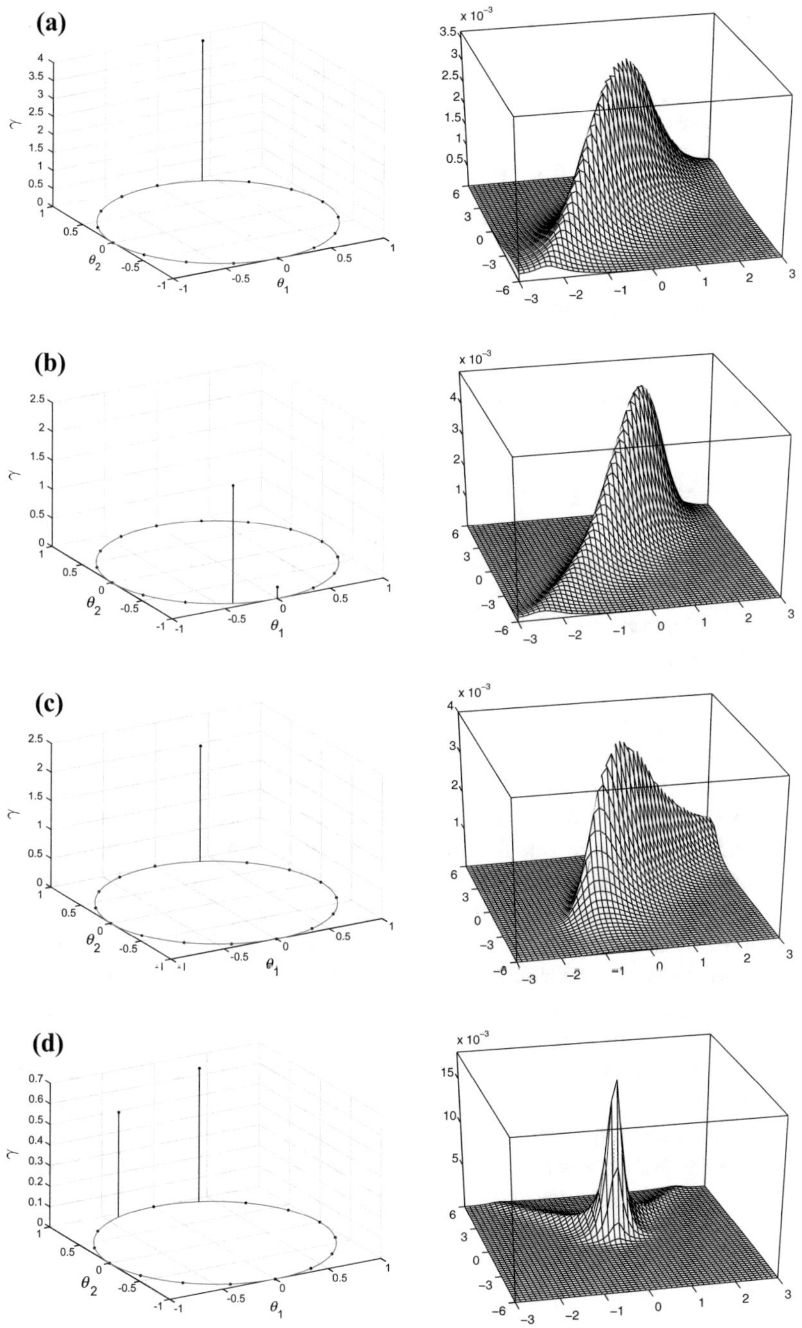

Fig. 6.2 Pseudo-optimal stable distributions $\alpha = 2$ (**a**), $\alpha = 1.5$ (**b**), $\alpha = 1$ (**c**), $\alpha = 0.5$ (**d**) for Rosenbrock's problem—on the left: weights for individual DSM focus points; on the right: corresponding two-dimensional probability density functions

Table 6.2 Evolutionary strategy $(1, \lambda/\nu)\text{ES}_\alpha$ with the DSM adaptation mechanism

Step 0: Set up λ, $\nu = \frac{\lambda}{2}$, $k = 1$. Choose initial approximation of the global solution x_0, the set of spinning vectors $\xi = \{s_i \mid i = 1, 2, \ldots, n_s\}, s_i \in S_n$ and the corresponding initial weight vector γ_k.
Step 1: Generate the set of alternative solutions using model $\Gamma_D(\xi, \gamma_k)$:

$$P_{k,\lambda} = \{x_{k,1}, x_{k,2}, \ldots, x_{k,\lambda}\}, \quad x_{k,i} = x_k + X_{\xi}^{\gamma_k}. \tag{6.15}$$

Step 2: Choose ν best solutions in the population $P_{k,\lambda}$:

$$P_{k,\nu} = \{x_{k,1:\lambda}, x_{k,2:\lambda}, \ldots, x_{k,\nu:\lambda}\}. \tag{6.16}$$

Step 3: Reconfigure the search model using the population $P_{k,\nu}$:

$$\gamma_k = E(P_{k,\nu}), \tag{6.17}$$

where $E(\cdot)$ is the procedure of estimation of weights γ_k of vectors spinning the spectral measure.
Step 4: Take $x_k = x_{k,1:\lambda}$.
Step 5: If the stop criterion is not met, go to Step 1.

operator. The evaluated sample solutions are subjected to selective elimination. The distribution of the remaining solutions is used to intuitive recombine the probabilistic model. The above procedure has been successively applied to solve problems of an engineering nature (Larranaga and Lozano 2001).

Stable distributions based on the DSM are defined by a considerable number of parameters. We have to determine the number, location and weights of specific vectors spinning the discrete spectral measure. Inclusion all of these factors in one adaptation procedure is a problem which needs a series of long-term studies. This will be done in the nearest future. Now, we assume a few simplifications, whose aim is to avoid too large calculation effort connected with model adaptation. The first assumption is that points which spin the DSM are uniformly located on the unit sphere surface. Thus, our problem is reduced to selection of proper weighs of previously chosen vectors. Because a large number of spinning vectors increases the computational effort of pseudo-random vector generation (6.6), the spinning vectors whose weights are lower than some relevance level should be ignored in the model.

The proposed optimization procedure (Obuchowicz and Prętki 2011; Prętki 2008) is based on the evolutionary strategy $(1, \lambda)$ES, which uses the adaptation schema presented in Table 6.2. Such a strategy will be denoted as $(1, \lambda/\nu)\text{ES}_\alpha$. The step which needs a more detailed comment is connected with the procedure of weight estimation for the vectors (6.17) which spin the DSM based on the distribution of the reduced solution population. This procedure is described in the next subsection.

6.3.1 DSM Estimation

The problem of discrete spectral measure estimation has been considered several times in the context of stochastic modelling (Chrysostomos and Shao 1981; Georgiou et al. 1999; Kidmose 2000; Nolan et al. 2001). There are a few methods of probability model design using this class of distributions (Nolan et al. 2001). One of the simplest ones, proposed by Nolan et al. (2001), will be applied for $(1, \lambda/\nu)\mathrm{ES}_\alpha$. This method is based on the so-called *empirical characteristic function*,

$$\hat{\varphi}(\boldsymbol{k}) = \frac{1}{N} \sum_{i=1}^{N} \exp(j\boldsymbol{k}^T \boldsymbol{X}_i), \tag{6.15}$$

where $\boldsymbol{X}_i$ are observed random values collected in the data set $\{\boldsymbol{x}_i\}_{i=1}^N$. Assuming that the discrete spectral measure Γ is defined by the finite set of vectors $\boldsymbol{\xi} = \{\boldsymbol{s}_i\}_{i=1}^{n_s}$ and the corresponding weights $\boldsymbol{\gamma} = \{\gamma_i\}_{i=1}^{n_s}$, i.e.,

$$\Gamma_s = \begin{Bmatrix} \boldsymbol{s}_1 \ \boldsymbol{s}_2 \ \dots \ \boldsymbol{s}_{n_s} \\ \gamma_1 \ \gamma_2 \ \dots \ \gamma_{n_s} \end{Bmatrix}, \tag{6.16}$$

the estimation problem can be reduced to the optimization task

$$\Gamma_s^* = \arg\min_{\boldsymbol{\xi},\boldsymbol{\gamma}} \|\hat{\varphi}(\boldsymbol{k}) - \varphi(\boldsymbol{k}; \boldsymbol{\xi}, \boldsymbol{\gamma})\|. \tag{6.17}$$

Because the DSM considered is spined on the fixed net of points uniformly distributed on the unit sphere surface $\boldsymbol{\xi} = \boldsymbol{\xi}_{n_s}$, the problem of model estimation (6.17) is reduced to the following form:

$$\boldsymbol{\gamma}^* = \arg\min_{\boldsymbol{\gamma}} \|\hat{\varphi}(\boldsymbol{k}) - \varphi(\boldsymbol{k}; \boldsymbol{\xi}_{n_s}, \boldsymbol{\gamma})\|. \tag{6.18}$$

The search for the exact solution of the problem (6.18) is connected with some serious computational effort and questions about the application of this method to determine the optimal set of weights for the spectral measure $\boldsymbol{\gamma}^*$ in the evolutionary algorithm. There is a need to introduce another simplification. The formula (6.18) is estimated using the set of testing points $\boldsymbol{K} = \{\boldsymbol{k}_i\}_{i=1}^{n_k}$ in the form

$$\boldsymbol{\gamma}^* = \arg\min_{\boldsymbol{\gamma}} \sum_{i=1}^{n_k} \Big(\hat{\varphi}(\boldsymbol{k}_i) - \varphi(\boldsymbol{k}_i; \boldsymbol{\xi}_{n_s}, \boldsymbol{\gamma})\Big)^2. \tag{6.19}$$

Let $\boldsymbol{I} = -[\ln \hat{\varphi}(\boldsymbol{k}_1), \dots, \ln \hat{\varphi}(\boldsymbol{k}_{n_k})]^T$ and

$$\boldsymbol{\Psi}(\boldsymbol{k}_1, \ldots, \boldsymbol{k}_{n_k}; \boldsymbol{s}_1, \ldots, \boldsymbol{s}_{n_s}) = \begin{pmatrix} \psi_\alpha(\boldsymbol{k}_1^T \boldsymbol{s}_1) & \ldots & \psi_\alpha(\boldsymbol{k}_1^T \boldsymbol{s}_{n_s}) \\ \vdots & \vdots & \vdots \\ \psi_\alpha(\boldsymbol{k}_{n_k}^T \boldsymbol{s}_1) & \ldots & \psi_\alpha(\boldsymbol{k}_{n_k}^T \boldsymbol{s}_{n_s}) \end{pmatrix} \tag{6.20}$$

for

$$\psi_\alpha(u) = \begin{cases} |u|^\alpha (1 - i \,\mathrm{sgn}(u) \tan(\frac{\pi\alpha}{2})), & \text{for } \alpha \neq 1, \\ |u|(1 - i \,\frac{2}{\pi} \mathrm{sgn}(u) \ln(|u|)), & \text{for } \alpha = 1. \end{cases} \tag{6.21}$$

Therefore, the optimal set of weights is the solution of the simultaneous equations (Nolan et al. 2001)

$$\boldsymbol{I} = \boldsymbol{\Psi}\boldsymbol{\gamma}^*. \tag{6.22}$$

In order to ensure a good condition of the problem (6.22), let us assume $n_s = n_k$ and $\boldsymbol{s}_i = \boldsymbol{k}_i$. Finally, the problem of optimal spectral measure weight designation has the form of a quadratic problem with inequality constrains (Nolan et al. 2001):

$$\boldsymbol{\gamma}^* = \arg\min_{\boldsymbol{\gamma} \geq \boldsymbol{0}} ||\boldsymbol{c} - \boldsymbol{A}\boldsymbol{\gamma}||_2, \tag{6.23}$$

where $\boldsymbol{c} = [\mathrm{Re}\{I_{1:n/2}\}, \mathrm{Im}\{I_{n/2+1:n}\}]^T$ is a vector containing firstly n real values of the vector I and next its n imaginary values, while $\boldsymbol{A} = [\mathrm{Re}\{\boldsymbol{\psi}_1^T\}, ..., \mathrm{Im}\{\boldsymbol{\psi}_n^T\}]^T$ is a similarly organized matrix (6.20), where $\boldsymbol{\psi}_i = [\psi(\boldsymbol{s}_1^T \boldsymbol{s}_i), \psi(\boldsymbol{s}_2^T \boldsymbol{s}_i), \ldots, \psi(\boldsymbol{s}_{n_s}^T \boldsymbol{s}_i)]^T$ are individual rows.

The problem (6.23) can be solved using one of the dedicated gradient methods.

6.3.2 Simulation Example

The thesis that optimization effectiveness of a given evolutionary strategy can be improved by DSM parameter adaptation, formulated at the beginning of this chapter, will be experimentally verified in this subsection.

Three versions of the evolutionary strategy will be analyzed:

- Version **A1**: the evolutionary strategy $(1, \lambda)\mathrm{ES}_\alpha$, for which the mutation vector is based on the DSM described by the vectors

$$\boldsymbol{\xi}_4 = \left\{ \begin{bmatrix} 1 \\ 0 \end{bmatrix}, \begin{bmatrix} 0 \\ 1 \end{bmatrix}, \begin{bmatrix} -1 \\ 0 \end{bmatrix}, \begin{bmatrix} 0 \\ -1 \end{bmatrix} \right\}.$$

The weight vector $\boldsymbol{\gamma} = [\sigma/4, \sigma/4, \sigma/4, \sigma/4]^T$ is fixed and remains unchanged during the time of the algorithm run.

- Version **A2**: the evolutionary strategy $(1, \lambda/\nu)\mathrm{ES}_\alpha$, for which the mutation vector is based on the DSM described by the vectors

$$\boldsymbol{\xi}_4 = \left\{ \begin{bmatrix} 1 \\ 0 \end{bmatrix}, \begin{bmatrix} 0 \\ 1 \end{bmatrix}, \begin{bmatrix} -1 \\ 0 \end{bmatrix}, \begin{bmatrix} 0 \\ -1 \end{bmatrix} \right\}.$$

  The initial weight vector $\boldsymbol{\gamma} = [\sigma/4, \sigma/4, \sigma/4, \sigma/4]^T$ is adapted using the algorithm (Table 6.2).
- Version **A3**: the evolutionary strategy $(1, \lambda/\nu)\mathrm{ES}_\alpha$, for which the mutation vector is based on the DSM described by the vectors

$$\boldsymbol{\xi}_8 = \left\{ \begin{bmatrix} 1 \\ 0 \end{bmatrix}, \begin{bmatrix} \frac{\sqrt{2}}{2} \\ \frac{\sqrt{2}}{2} \end{bmatrix}, \begin{bmatrix} 0 \\ 1 \end{bmatrix}, \begin{bmatrix} -\frac{\sqrt{2}}{2} \\ \frac{\sqrt{2}}{2} \end{bmatrix}, \begin{bmatrix} -1 \\ 0 \end{bmatrix}, \begin{bmatrix} -\frac{\sqrt{2}}{2} \\ -\frac{\sqrt{2}}{2} \end{bmatrix}, \begin{bmatrix} 0 \\ -1 \end{bmatrix}, \begin{bmatrix} -\frac{\sqrt{2}}{2} \\ -\frac{\sqrt{2}}{2} \end{bmatrix} \right\}.$$

  The initial weight vector $\boldsymbol{\gamma} = [\sigma/8, \ldots, \sigma/8]^T$ is adapted using the algorithm (Table 6.2).

It is worth noting that mutation in the version **A1** is the same as in the case of the mutation vector composed of two independent random variables of a spherical symmetry: $\mathbf{Z}[S\alpha S(\sigma/2), S\alpha S(\sigma/2)]^T$. The algorithm of the version **A2** has the mutation operator based on the vector composed of two independent random variables of the stable distribution with slope parameters different than zero, $\mathbf{Z}[S_\alpha(\sigma/2, \beta_1),]^T$.

Four unimodal functions defining ill-conditioned local optimization problems are used in the experiment in order to compare the above-described strategies,

$$f(x_1, x_2) = [x_1, x_2]^T \begin{pmatrix} -\frac{\sqrt{2}}{2} & \frac{\sqrt{2}}{2} \\ \frac{\sqrt{2}}{2} & \frac{\sqrt{2}}{2} \end{pmatrix} \begin{pmatrix} e_1 & 0 \\ 0 & e_2 \end{pmatrix} \begin{pmatrix} -\frac{\sqrt{2}}{2} & \frac{\sqrt{2}}{2} \\ \frac{\sqrt{2}}{2} & \frac{\sqrt{2}}{2} \end{pmatrix}^T [x_1, x_2], \tag{6.24}$$

for different condition factors (the ratio of the biggest eigenvalue to the smallest one): $(e_1, e_2) = (1, 1), (10, 0.1), (100, 0.01), (1000, 0.001)$. Initial conditions for all algorithms are the same: $\lambda = 20$, $\nu = 10$, $\boldsymbol{x}_0 = [1000, 1000]^T$, $\sigma = 1$. Moreover, in order to emphasise the estimation role of preferred mutation directions, the following constrain on DSM weights are applied: $\sum_{i=1}^{n_s} \gamma_i = \sigma$. This is a way to prevent increasing the entropy of mutation random vectors. Results in the form of averaged optimization processes of algorithms with the stability index $\alpha = 0.5$ are presented in Fig. 6.3.

Analyzing the results in Fig. 6.3, the advantage of the adaptation mechanism applied cannot be clearly established. This is undoubtedly caused by the chosen stability index $\alpha = 0.5$, for which macro-mutations are the main cause of relatively quick extremum localization. Although the algorithm **A1** has substantial problems with selection of the most beneficial direction of mutation for the functions considered (the mutation operator in this algorithm prefers directions parallel to axis of the reference frame—the so-called symmetry effect described by Obuchowicz (2003b) and in Sect. 4.2), the macro-mutation mechanism can effectively locate an

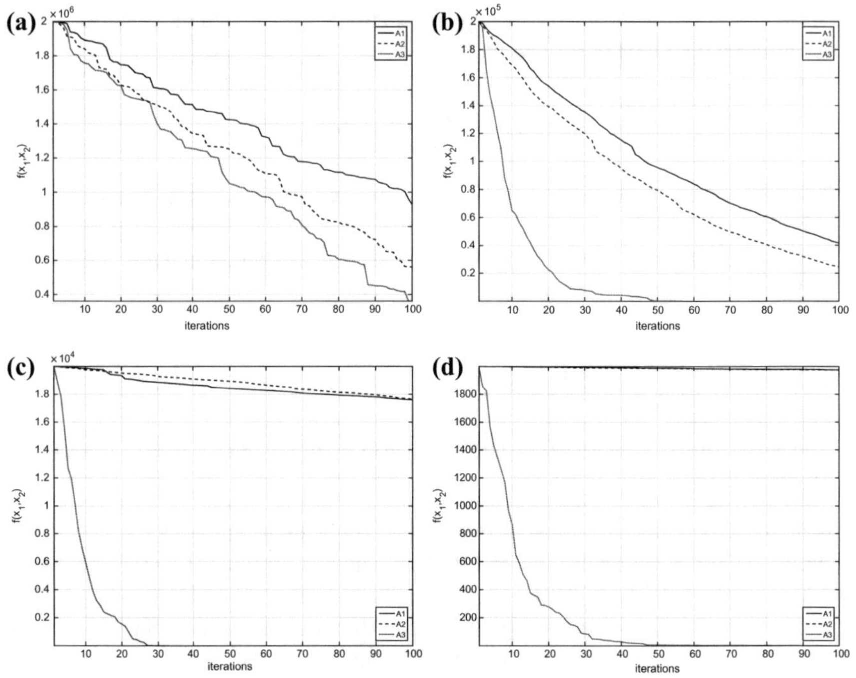

Fig. 6.3 Averaged process of quadratic function optimization using algorithms with different discrete spectral measures: **A1**—four-point DSM without adaptation, **A2**—four-point DSM with adaptation, **A3**—eight-point DSM with adaptation. Results are averaged over 50 independent algorithm runs for each evolutionary strategy. Parameters of the objective function (6.24) (e_1, e_2): (1, 1) (**a**), (10, 0.1) (**b**), (100, 0.01) (**c**), (1000, 0.001) (**d**)

extremum in few subsequent iterations. The algorithm **A2** differs from the one above in the possibility to adapt parameters of the mutation distribution. It causes better optimization effectiveness, but has a problem with fitting the exploration distribution to any direction of the searching space. Such a possibility occurs in the case of the algorithm **A3** thanks to a more dense configuration of vectors spanning the DSM. Observing the trajectory presented in Figs. 6.4b–d, it easy to see how the relatively largest value of the weight $\gamma_6 \approx \sigma$ of the vector $s_6 = [-\frac{\sqrt{2}}{2}, -\frac{\sqrt{2}}{2}]^T$ influencess the algorithm. Such configuration of the DSM causes potential macro-mutations to occur only in the direction parallel to s_6. This results in incomparably better convergence of the algorithm **A3** in the case of ill-conditioned testing problems (Fig. 6.3).

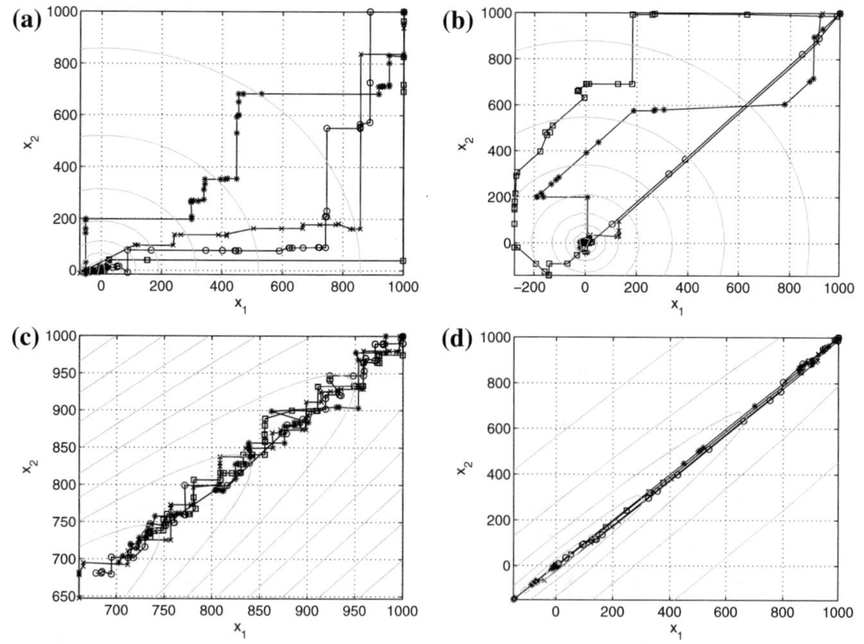

Fig. 6.4 Sample trajectories of the best solutions of the algorithms **A1**: (**a**, **c**) and **A3**: (**b**, **d**). Figures (**a**, **b**) correspond to objective functions with the parameters $(e_1, e_2) = (1, 1)$, while (**c**, **d**) to $(e_1, e_2) = (10, 0.1)$

6.4 Summary

Until now, application of multidimensional stable distributions in global optimization algorithms has been limited to the simplest cases: the $NS\alpha S(\sigma)$ distribution (Lee and Yao 1994; Obuchowicz 2003b; Obuchowicz and Prętki 2004; Yao and Liu 1997; Yao and Liu 1999) or the IS distribution (Obuchowicz and Prętki 2005; Prętki and Obuchowicz 2008; Rudolph 1997). DSM application allows generating a wide class of random vectors and modelling complicated stochastic relations between decision variables. We access many stable distribution properties, which may turn out very valuable in the context of stochastic optimization. A method of DSM estimation for the standard evolutionary strategy $(1+1)\text{ES}_\alpha$ was presented in this chapter. The multidimensional stable model, used for searching space exploration, is fitted to the environment based on information covered in the distribution of the best elements in the population. A series of simple experiments clearly shows the advantage of this method over proposals which do not have the possibility to take into account strong correlations between decision variables.

The proposed method has some disadvantages. DSM weight estimation needs the expensive local optimization process. Moreover, it is desirable that the number of vectors spanning the DSM sufficiently increase with the space dimension increasing so as to allow detecting complex relations between individual decision variables. In the case of the lack of any a priori knowledge about an optimization problem, this number, as well as calculation cost of DSM estimation, increases exponentially. A solution to this issue should be looked for using heuristic methods for selection of spanning vectors and adaptation of their weights. This is the aim of our future research.

References

Byczkowski, T., Nolan, J. P., & Rajput, B. (1993). Approximation of multidimensional stable densities. *Journal of Multivariate Analysis*, *46*, 13–31.

Chrysostomos, L. N., & Shao, M. (1981). *Signal processing with alpha-stable distribution and applications*. Chichester: Wiley.

Durrett, R. (1995). *Probability: Theory and examples*. Belmont: Duxbury Press.

Georgiou, P. G., Tsakalides, P., & Kyriakakis, C. (1999). Alpha-stable modeling of noise and robust time-delay estimation in the presence of impulsive noise. *IEEE Transactions on Multimedia*, *1*(3), 291–301.

Kemp, F. (2003). An introduction to sequential Monte Carlo methods. *Journal of the Royal Statistical Society: Series D*, *52*, 694–695.

Kidmose, P. (2000). Alpha-stable distributions in signal processing of audio signals. In *41st Conference on Simulation and Modelling* (pp. 87–97). Scandinavian Simulation Society Press.

Larranaga, P., & Lozano, J. A. (2001). *Estimation of distribution algorithms: A new tool for evolutionary optimization*. Boston: Kluwer Academic.

Lee, C. Y., & Yao, X. (1994). Evolutionary programming using mutation based on the Lévy probability distribution. *IEEE Transactions on Evolutionary Computation*, *9*(1), 1–13.

MacKey, D. C. J. (1998). Introduction to the Monte Carlo methods. In M. I. Jordan (Ed.), *Learning in graphical models* (pp. 175–204). NATO science series. Dordrecht: Kluwer Academic Press.

Modarres, R., & Nolan, J. P. (2002). A method for simulating stable random vectors. *Computational Statistics*, *9*, 11–19.

Nolan, J. P., Panorska, A. K., & McCulloch, J. H. (2001). Estimation of stable spectral measures-stable non-Gaussian models in finance and econometrics. *Mathematical and Computer Modelling*, *34*(9), 1113–1122.

Obuchowicz, A. (2003). *Evolutionary algorithms in global optimization and dynamic system diagnosis*. Zielona Góra: Lubuskie Scientific Society.

Obuchowicz, A., & Prętki, P. (2004). Phenotypic evolution with mutation based on symmetric α-stable distributions. *International Journal on Applied Mathematics and Computer Science*, *14*(3), 289–316.

Obuchowicz, A., & Prętki, P. (2005). Isotropic symmetric α-stable mutations for evolutionary algorithms. In I. E. E. E. Congress (Ed.), *on Evolutionary Computation* (pp. 404–410). UK: Edinbourgh.

Obuchowicz, A., & Prętki, P. (2010). Evolutionary algorithms with stable mutations based on a discrete spectral measure. In L. Rutkowski, R. Scherer, R. Tadeusiewicz, L. A. Zadeh, & J. M. Zurada (Eds.), *Artificial intelligence and soft computing: Part II* (Vol. 6114, pp. 181–188). Lecture notes on artificial intelligence. Berlin: Springer.

Obuchowicz, A., Prętki, P. (2011). Self-adapted stable mutation based on discrete spectral measure for evolutionary algorithms. In *13th Conference on Evolutionary Algorithms and Global Optimization*. Warsaw: Warsaw University of Technology Press.

Prętki, P. (2008). *α-Stable distributions in evolutionary algorithms for global parametric optimization*. Ph.D. thesis, University of Zielona Góra (in Polish).

Prętki, P., & Obuchowicz, A. (2008). Robustness of isotropic stable mutations in general search space. In L. Rutkowski, R. Scherer, R. Tadeusiewicz, L. A. Zadeh, & J. M. Zurada (Eds.), *Artificial intelligence and soft computing* (Vol. 5097, pp. 460–468). Lecture notes on Artificial Intelligence. Berlin: Springer.

Rudolph, G. (1997). Local convergence rates of simple evolutionary algorithms with Cauchy mutations. *IEEE Transactions on Evolutionary Computation*, *1*(4), 249–258.

Samorodnitsky, G., & Taqqu, M. S. (1994). *Stable non-Gaussian random processes*. New York: Chapman and Hall.

Spall, J. C. (1998). Implementation of simultaneous perturbation algorithm for stochastic optimization. *IEEE Transactions on Aerospace and Electronic Systems*, *34*, 817–823.

Spall, J. C. (2003). *Introduction to stochastic search and optimization*. Hoboken: Wiley.

Vidyasagar, M. (2001). Randomized algorithms for robust controller synthesis using statistical learning theory. *Automatica*, *37*(10), 1515–1528.

Yao, X., & Liu, Y. (1996). Fast evolutionary strategies. Control. *Cybernetics*, *26*(3), 467–496.

Yao, X., & Liu, Y. (1999). Evolutionary programming made faster. *IEEE Transactions on Evolutionary Computation*, *3*(2), 82–102.

Chapter 7
Isotropic Mutation Based on an α-Stable Generator

Application of the isotropic distribution based on an α-stable generator $S\alpha SU(\sigma)$ (3.57) to a mutation operator in evolutionary algorithms seems to be closest to our purpose, which is the elimination of the dead surrounding effect. This conclusion stems from the fact that the random variable $\|X\|$, where $X \sim S\alpha SU(\sigma)$, depends only on the generating variable $\|R\|$ and not on the space dimension n (or a proof, see (3.59)). Also, we hope that this fact weakens the dead surrounding effect, so important in the case of $NS\alpha S(\boldsymbol{\sigma})$ and $IS\alpha S(\sigma)$. Can it be eliminated completely? We try to answer this question using an experiment described in the first section. Next, like in Chap. 5, the local convergence analysis of the$(1+1)\text{ES}_{S;\alpha}$ and $(1+\lambda)\text{ES}_{S;\alpha}$ strategies, as well as simulation research on exploration and exploitation abilities of the $\text{ESTS}_{S;\alpha}$ algorithms, is presented. Finally simulation analysis of the $\text{ESSS}_{S;\alpha}$ algorithms' effectiveness in chosen global optimization problems is presented.

7.1 Dead Surrounding Effect

In order to analyze the dead surrounding effect, an experiment is performed. Namely, the sphere function $f_{sph}(\boldsymbol{x})$ (B.1) is chosen as an objective one. The initial population composed of $\eta = 20$ elements is generated by 20 mutations of the extremum point $\boldsymbol{x}_0^0 = [0, \ldots, 0]$. Two pairs of algorithms are used in the experiment: one is evolutionary search with soft selection and the non-isotropic stable mutation $NS\alpha S(\boldsymbol{\sigma})$ $\text{ESSS}_{N,\alpha}$ for $\alpha = 0.5$ and $\alpha = 1$, while the other is composed of the same evolutionary algorithm $\text{ESSS}_{S,\alpha}$, but with the isotropic mutation $S\alpha SU(\sigma)$ based on an α-stable generator and the same stability indices α. The scale parameter $\sigma = 0.05$ is chosen the same for all algorithms. Each algorithm is started 100 times for each set of initial parameters. The maximal number of iterations is chosen as $t_{\max} = 10000$.

A. Obuchowicz, *Stable Mutations for Evolutionary Algorithms*,
Studies in Computational Intelligence 797,
https://doi.org/10.1007/978-3-030-01548-0_7

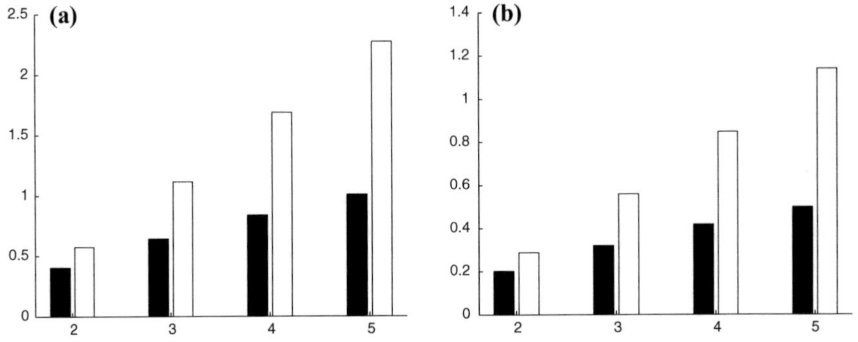

Fig. 7.1 Mean distance between the extremum and the best elements in the population fluctuated around it versus the space dimension. Results obtained for $\text{ESSS}_{N,\alpha}$ (white bars) and $\text{ESSS}_{S\alpha}$ (black bars) with the stability indices $\alpha = 1$ (**a**) and $\alpha = 0.5$ (**b**)

Figure 7.1 illustrates at what distance from the extremum the fluctuated population is stabilized for the different mutations applied.

In the case of $\text{ESSS}_{N,\alpha}$ algorithms with non-isotropic stable mutation, population fluctuations are more distant from the extremum with the space dimension increasing, which is in accordance with the results presented in Sect. 4.1 (e.g., Fig. 4.1). What is more surprising, this effect is visible, although significantly weakened, despite isotropic mutation based on the application of the α-stable generator $S\alpha SU(\sigma)$.

In order to explain the lack of possibility of full elimination of the dead surrounding effect in the above-described experiment, the *probability of successful mutation* ζ should be introduced. ζ describes the probability of offspring location in an area of better values of the objective function in comparison to its parent. Such a probability, in the case of the sphere function $f_{sph}(\boldsymbol{x})$ (B.1), can be defined as follows:

$$\zeta(\boldsymbol{x}) = \int_{\mathscr{K}(\boldsymbol{x})} p(\boldsymbol{x})d\boldsymbol{x} \quad \text{where} \quad \mathscr{K}(\boldsymbol{x}) = \{\boldsymbol{y} : \|\boldsymbol{y}\| < \|\boldsymbol{x}\|\}, \tag{7.1}$$

where $p(\cdot)$ is the probability density function of the mutation distribution. In the case of $S\alpha SU(\sigma)$, the function $p(\cdot)$ is described by (3.63). It is worth noting that the integration area $\mathscr{K}(\boldsymbol{x})$ decreases with $\|\boldsymbol{x}\| \to 0$. Figure 7.2 presents a numerically calculated relation $\zeta(\|\boldsymbol{x}\|)$ for the $NS\alpha S(\boldsymbol{\sigma})$ and $S\alpha SU(\sigma)$ distributions, the stability indices $\alpha = 1$ and $\alpha = 2$ as well as the scale parameter $\sigma = 0.05$.

The probability of successful mutation decreases with approaching the extremum, both in the case of $NS\alpha S(\boldsymbol{\sigma})$ and $S\alpha SU(\sigma)$ (Fig. 7.2). However, the dependence between this probability and the space dimension in the case of the $S\alpha SU(\sigma)$ distribution is not as strong as in the case of the $NS\alpha S(\boldsymbol{\sigma})$ distribution.

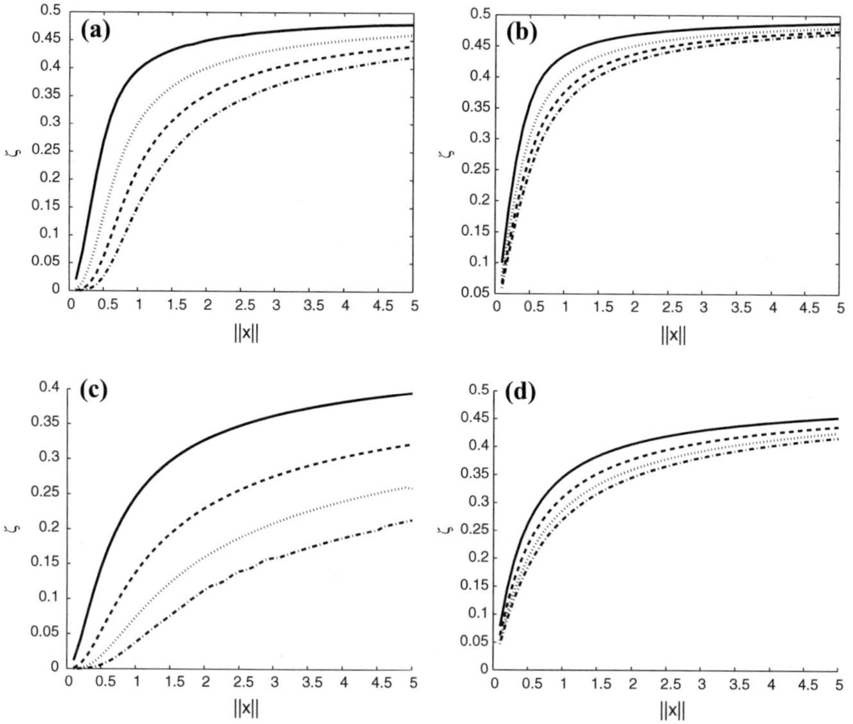

Fig. 7.2 Probability of successful mutation ζ for the spherical function and mutation distributions: $NS\alpha S(\boldsymbol{\sigma})$ for $\alpha = 2$ (**a**), $S\alpha SU(\sigma)$ for $\alpha = 2$ (**b**), $NS\alpha S(\boldsymbol{\sigma})$ for $\alpha = 1$ (**c**), $S\alpha SU(\sigma)$ for $\alpha = 1$ (**d**) versus the distance between the base point and the extremum ($n = 2$ (solid line), $n = 3$ (dotted line), $n = 4$ (dashed line), $n = 5$ (dash-dot line))

7.2 Local Convergence Analysis

The analysis of the local convergence of an evolutionary algorithm with isotropic mutation based on an α-stable generator will be performed in the same way as in the case of isotropic stable mutation (Sect. 5.2).

7.2.1 Local Convergence of the $(1 + 1)ES_{S,\alpha}$ Strategies

First let us consider the sphere function $f_{sph}(\boldsymbol{x})$ (B.1) as an objective one and the class of evolutionary strategies $(1 + 1)\text{ES}_{S,\alpha}$. The only element which distinguishes the current discussion from that described in Sect. 5.2 is the form of the probability density function $p_V(v, \alpha)$ of the random variable V (5.14) in the formula (5.16), where the expectation value of the progress rate is described. This density can be calculated (Rudolph 1997) using definition of the probability density function of the

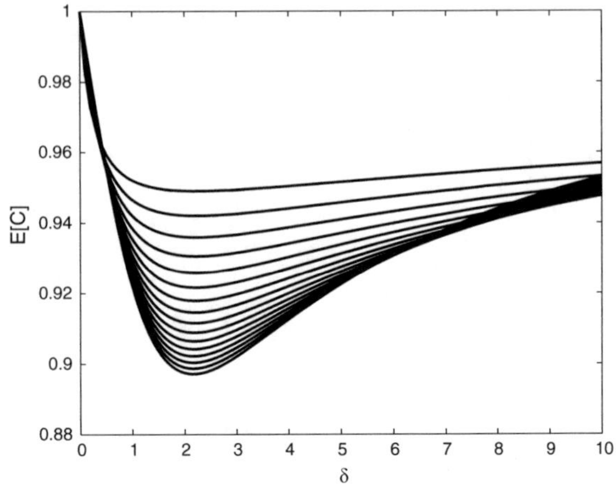

Fig. 7.3 Expectation value of the progress rate of the $(1+1)ES_{S,\alpha}$ strategies versus δ. Lines from top to bottom correspond to subsequent values of the stability index α from 0.5 to 2.0 with step 0.1

distribution $S\alpha SU(\sigma)$ (3.63) and can written in the form (Obuchowicz and Prętki 2005)

$$p_V(v,\alpha) = \frac{\delta v^{n/2-1}}{B\left(\frac{n-1}{2},\frac{1}{2}\right)} \int_{-1}^{1} \frac{p_{\alpha,1}\left(\delta(v-2t\sqrt{v}+1)^{1/2}\right)}{(v-2t\sqrt{v}+1)^{(n-1)/2}} (1-t^2)^{(n-3)/2} dt, \quad (7.2)$$

where, like in (5.17), $\delta = \frac{\|\boldsymbol{x}_k\|}{\sigma}$ is the length of the vector $\boldsymbol{x}_k$ measured in the scale parameter units, $B(\cdot,\cdot)$ is the Beta function and $p_{\alpha,\sigma}(\cdot)$ is the probability density function of the random variable $R \sim S\alpha S(\sigma)$. Unfortunately, the expectation value of the progress rate can be calculated only numerically, because of the lack of knowledge about the analytical form of the function $p_{\alpha,1}(\cdot)$. Results of numerical calculations for $\alpha = 0.5, 0.6, \ldots, 2$ and $\delta = 0.1, 0.2, \ldots, 10$ are presented in Fig. 7.3.

It is interesting that the localization of the progress rate extremum does not depend on the stability index α and is reached for $\delta^\star \approx 2.2$. Mutations with heavy tails, in general, are connected with slower convergence to a local extremum. It is worth noting that the range of δ for all values of α can be divided into two partial ranges: $(0, \delta^\star]$ and $(\delta^\star, \infty)$. A relevant difference between them is the following: the effectiveness of the $(1+1)ES_{S,\alpha}$ strategy is more sensitive to possible changes of δ in $(0, \delta^\star]$ than to the same changes in the second range. Relatively deeper minimums of the relation between the progress rate and δ for high values of α can lead to the conclusion that an appropriate choice of the optimal values of the scale parameter $\sigma^\star = \|\boldsymbol{x}_k\|/\delta^\star$ can be much harder for these stability indices. In the case of low values of α, the optimal choice of σ is not as crucial a problem.

7.2.2 Local Convergence of the $(1+\lambda)ES_{S,\alpha}$ *Strategies*

In the evolutionary strategy $(1+\lambda)ES_{S,\alpha}$, λ offspring are generated from the parent element $\boldsymbol{x}_k$ in iteration k using the mutation operator $\boldsymbol{x}'_{k,i} = \boldsymbol{y}_k + X_i$, where

$(X_i \mid i = 1, 2, \ldots, \lambda)$ is a sequence of independent random vectors of the $S\alpha SU(\sigma)$ distribution. Only the best element from the set $\{\boldsymbol{x}_k, \boldsymbol{x}'_{k,1}, \ldots, \boldsymbol{x}'_{k,\lambda}\}$ is kept in the next generation.

Another statistical theory (Kim et al. 2002) will be applied in order to estimate the progress rate. Let X be a random variable with the probability density function $p(x)$ and cumulative distribution function $F(x)$. Let the sequence $S_\lambda = (X_1, X_2, \ldots, X_\lambda)$ denote independent samples of X with the same distribution. Let us to transform S_λ to the form of ordered statistics, i.e., $X_{1:\lambda} \le X_{2:\lambda} \le \cdots \le X_{\lambda:\lambda}$; then the probability density function of $X_{i:\lambda}$ is in the form

$$p_{X_{i:\lambda}}(x) = \frac{\lambda!}{(i-1)!(\lambda-i)!}\Big[F(x)\Big]^{i-1}\Big[1-F(x)\Big]^{\lambda-i} p(x). \tag{7.3}$$

It is easy to prove (Kim et al. 2002) that the progress rate of the $(1+\lambda)ES_{S,\alpha}$ strategy has the following form:

$$\varphi = E\Big[\min\Big\{V_{1:\lambda}, 1\Big\}\,\Big|\,||\boldsymbol{y}_k||\Big] = \int_0^1 (1-F_v(v))^\lambda dv. \tag{7.4}$$

Because the explicit form of (7.4) is unknown, the progress rate for $\delta = 2.2$ can be calculated numerically only, and the results are presented in Fig. 7.4.

The local convergence of the evolutionary strategy $(1+\lambda)\text{ES}_{S,\alpha}$ can be significantly accelerated by increasing the number of offspring λ. Taking into account the fact that the increasing number of elements in the population is connected with the increasing computational effort of algorithm simulation, one should work out some compromise between the progress rate expressed in the number of iterations and the number of fitness function evaluations in each iteration.

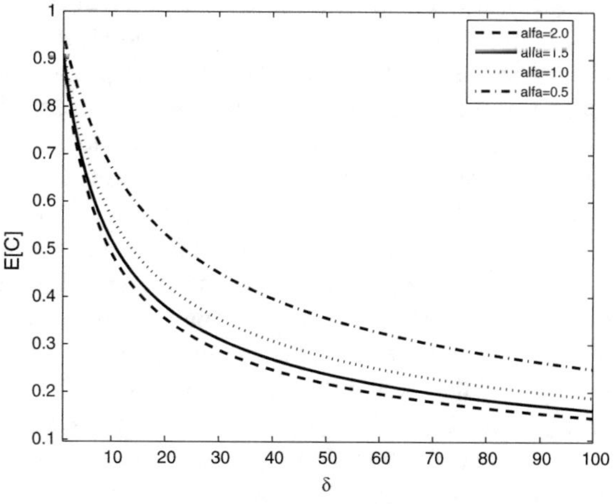

Fig. 7.4 Optimal values of the progress rate of the $(1+\lambda)ES_{S,\alpha}$ strategy vs. the number of offspring λ (lines from top to bottom correspond to the following values of the stability index: $\alpha = 0.5, 1, 1.5, 2$)

7.3 Exploitation and Exploration: ESTS$_{S,\alpha}$

Due to trivial hard selection in the simple evolutionary strategy considered in the previous subsection, a mathematical model of the evolutionary process can be created. An algorithm with soft selection will be analyzed here. Creation of mathematical models of an evolutionary process for algorithms with soft selection is much harder, thus there are a few research reports (Karcz-Dulęba 2001, 2004) with some additional restrictive conditions. Moreover, in this book we consider mutations which do not have explicit forms of the probability density function, thus looking for an analytical model is rather doomed to failure. Because there is no possibility of analytical parameter calculation (e.g., progress rate), we use parameters based on some intuitive approach: the mean distance of a fluctuated population from a local extremum and the mean number of iterations needed to cross a saddle between quality peaks of a fitness function.

7.3.1 Precision of Local Extremum Localization

Let us assume that the precision of optimal point localization is an important exploitation parameter. This approach is quite different than in the case of determination of the progress rate, which represents the rate of an algorithm's convergence to the local extremum.

The answer to the question regarding the ability of local extremum localization by an evolutionary algorithm with soft selection and mutation using stable heavy-tail distributions can be based on Theorem 3.11, proved in Sect. 3.1.4. The general property of all evolutionary algorithms with soft selection is that the population in the generation $k+1$ is usually generated from some subset of the population in the generation k. This means that some elements of this population are mutated several times and give more than one offspring, and, simultaneously, there is a subset of individuals which do not have any offspring in the next population. Let $\overline{r}$ be the mean number of parent element reproductions; in other words, on average, $\overline{r}$ elements of the $k+1$th population have the same parent. Based on Theorem 3.11, one can conclude that the expected localization of the best offspring of a given parent element will be much closer to the extremum than its parent if only the appropriate reproduction number $\overline{r}$ is guaranteed in an evolutionary algorithm.

The ESTS$_{S,\alpha}$ algorithm, as has been mentioned several times, does not localize exactly a local extremum, but its population fluctuates around the extremum at some distance. In order to estimate the distance $H(\sigma)$ between the best element in the current population and the extremum in the selection–mutation equilibrium state, an experiment is performed. Namely, the sphere function $f_{sph}(\boldsymbol{x})$ (B.1) is chosen as an objective one. The following parameters, which control the evolutionary process, are chosen: the population size $\eta = 20$, stability indices $\alpha = 0.5, 1, 1.5, 2$, scale parameters $\sigma = 0.001, \ldots, 0.009, 0.01, \ldots, 0.09, 0.1, \ldots, 0.5$, the tournament size

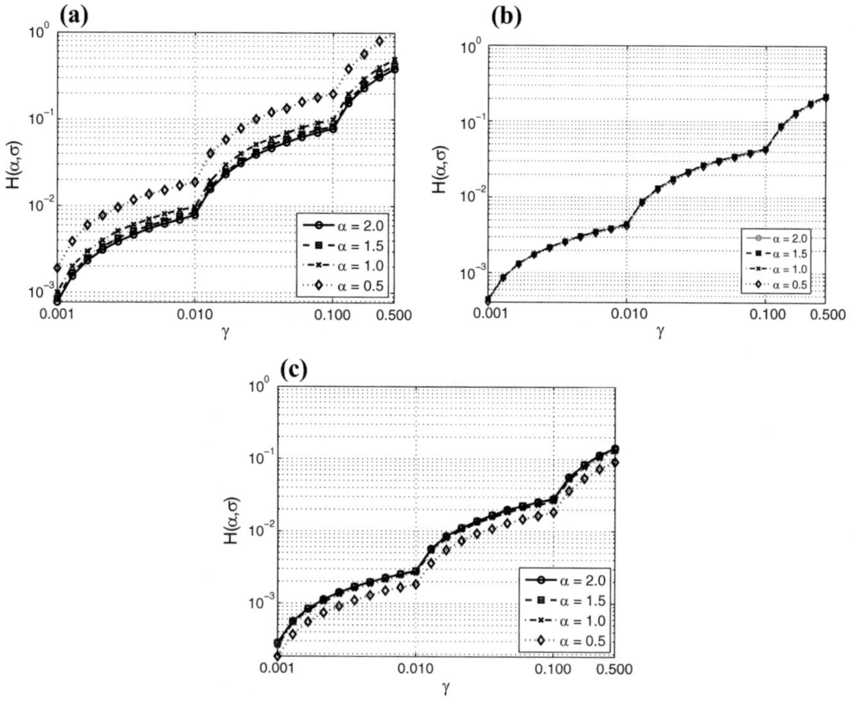

Fig. 7.5 Mean distances between the best elements and the extremum of the sphere function in the population equilibrium state. Tournament sizes: 2 (**a**), 4 (**b**), 8 (**c**)

$\eta_G = 2, 4, 8$, the space dimension $n = 4$. The results are presented in Fig. 7.5. Each point on the graphs is determined based on 10000 algorithm runs.

It is interesting that the relation between the distance $H(\alpha, \sigma)$ and the scale parameter σ is almost linear. The relation between $H(\alpha, \sigma)$ and the stability index is more complicated. For low values of the tournament size, the best results are obtained for $\mathrm{ESTS}_{S,2}$, while algorithms with mutation based on heavy-tail distributions localize the extremum much worse. This tendency is reversed when increasing the tournament size η_G. For $\eta_G = 4$, relationships $H(\alpha, \sigma)$ are almost the same. These results coincide with those presented in Fig. 3.2, where the expectation value of the dependence of the ordered random variable $X_{1:\lambda}$ on the number of offspring of the same parent is presented. As can be observed in Fig. 3.2, $\lambda = 4$ is also the border of shift of the relationship between the expectation value of $X_{1:\lambda}$ and α. This coincidence results from the fact that each element of the population is approximately η_G times selected for tournament groups in a given iteration by the tournament selection process in the $\mathrm{ESTS}_{S,\alpha}$ algorithm. Thus, the best element in the current population, winning in its tournament groups, gives approximately η_G of its copies to the indirect population.

7.3.2 Saddle Crossing

The number of iterations needed to cross a saddle between quality peaks of the problem defined in Appendix A can be chosen as some measure of the exploration ability of an evolutionary algorithm. The same evolutionary algorithm as in the previous subsection, i.e., $\mathrm{ESTS}_{S,\alpha}$, is used in the experiment, and the same values of control parameters are chosen: the population size $\eta = 20$, the stability index $\alpha = 0.5, 1, 1.5, 2$, the scale parameter $\sigma = 0.001, \ldots, 0.009, 0.01, \ldots, 0.09, 0.1, \ldots, 0.5$, the tournament size $\eta_G = 2, 4, 8$, the space dimension $n = 4$. The results are presented in Fig. 7.6. Each point on the graphs is obtained over 10000 algorithm runs.

The following fact is important: sensitivity of the saddle crossing effectiveness of the $\mathrm{ESTS}_{S,\alpha}$ algorithm to the scale parameter σ decreases with the stability index α decreasing. Moreover, the population crosses the saddle quicker for low values of α. In the case of a more diverse population ($\eta_G = 2$ (Fig. 7.6a)) for $\alpha = 0.5$ and $\sigma > 0.2$, as well as for $\alpha = 1, \sigma > 0.4$, a rapid decrease in this efficacy can be noticed. This means that when some limit of the competitive balance between mutation described by the pair (α, σ) and the selection pressure is undermined, then

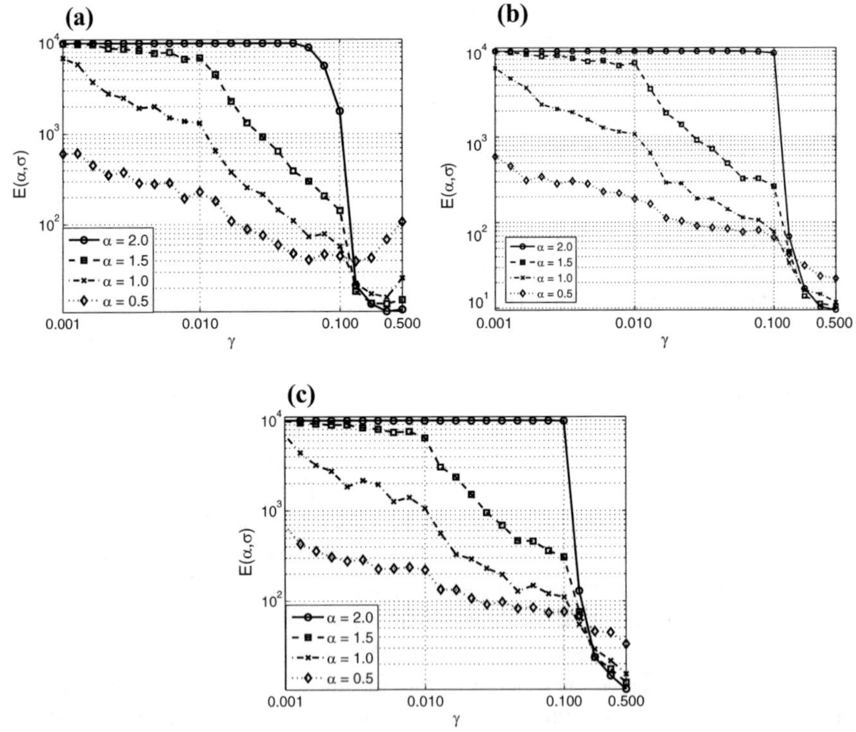

Fig. 7.6 Mean number of iterations E(α, σ) needed to cross the saddle between two Gaussian peaks (Appendix A). The tournament size: 2 (**a**), 4 (**b**), 8 (**c**)

the evolutionary process is more chaotic. This limit moves to increasing values of σ or decreasing values of α when the selection pressure increases. For a detailed discussion of this phenomenon, a precise detailed analytical model of this process should be formulated.

7.3.3 Exploitation Contra Exploration: The Pareto Front

Let us summarize the results obtained in both of the previous subsections. First of all, it can be noticed that exploitation abilities in phenotype evolutionary algorithms increase when the selection pressure increases (i.e., the tournament size η_G increases). On the other hand, it has been shown that the population controlled by lower stability indices α possesses promising exploration abilities, which are less sensitive to the scale parameter σ and the selection pressure η_G. One can ask whether, for a given η_G, a choice of parameters $\{\alpha, \sigma, \}$ which guarantees both high exploration and exploitation abilities is possible. The answer will be looked for by analyzing some quality factor of the $\mathrm{ESTS}_{S,\alpha}$ algorithm, which includes both criterions described above. Because none of the analyzed abilities can be directly compared, the idea of Pareto optimality, known from multiobjective optimization theory, will be applied. Let us define the vector quality factor $\boldsymbol{F} = [H(\alpha, \sigma), E(\alpha, \sigma)]^T$, where $H(\alpha, \sigma)$ denotes an estimation of the dead surrounding obtained using numerical simulations (see Sect. 7.3.1) and $E(\alpha, \sigma)$ is the number of iterations needed to cross a saddle between Gaussian quality peaks (see Sect. 7.3.2). Based on the Pareto optimality concept, we will say that the algorithm $\mathrm{ESTS}_{S,\alpha}$ with parameters $\{\alpha^\star, \sigma^\star\}$ is p-optimal if

$$\forall\{\alpha, \sigma\} \;\; \mathrm{H}(\alpha, \sigma) \geq \mathrm{H}(\alpha^\star, \sigma^\star) \text{ and } \mathrm{E}(\alpha, \sigma) \geq \mathrm{E}(\alpha^\star, \sigma^\star). \tag{7.5}$$

Relations between $\mathrm{H}(\alpha, \sigma)$ and $\mathrm{E}(\alpha, \sigma)$ for the evolutionary search $\mathrm{ESTS}_{S,\alpha}$ are presented in Fig. 7.7. It can be observed that mutation based on the $S\alpha SU(\sigma)$ distributions with low values of the stability index α allows a more effective compromise between exploitation and exploration abilities. Evolutionary search with these mutations dominates in a large part of the Pareto front, especially in the case of low values of the scale parameter σ. $\mathrm{ESTS}_{S,2}$ is becoming irreplaceable when the limit of the selection–mutation chaos for high values of σ and low values of α, which is mentioned at the end of Sect. 7.3.2, is overstepped.

7.4 Global Optimization

Let us consider 5D multi-modal Ackley $f_A(\boldsymbol{x})$ (B.3) and Rastrigin $f_R(\boldsymbol{x})$ (B.4) functions. In order to illustrate the effectiveness of mutation based on the $S\alpha SU(\sigma)$ distributions, algorithms of the class $\mathrm{ESSS}_{S,\alpha}$ are applied. The experiment is defined by the

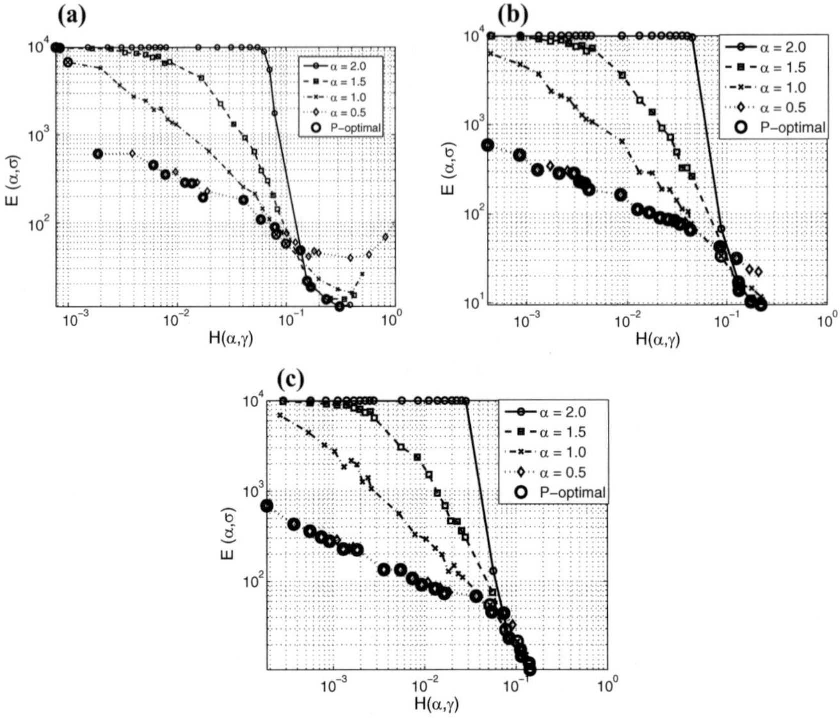

Fig. 7.7 Relations between $\mathrm{H}(\alpha, \sigma)$ and $\mathrm{E}(\alpha, \sigma)$ for the evolutionary search $\mathrm{ESTS}_{S,\alpha}$. Tournament size: 2 (**a**), 4 (**b**), 8 (**c**). Points which belong to the Pareto front are emphasised by big circles

following values of control parameters: the population size $\eta = 20$, the initial population is obtained by $\eta = 20$ mutations of the starting point $\boldsymbol{x}_0 = [10, 10, 10, 10, 10]$, the maximal number of iterations $t_{max} = 10000$, scale parameters are chosen from the interval $\sigma \in [0.0005; 1]$, stability indices $\alpha = 0.5, 1, 1.5, 2$. We recognize that the global extremum is reached if the objective function value of the best element is lower than 0.1 (Ackley's function) and 4.5 (Rastringin's function).

Figure 7.8 presents experiment results averaged over 100 algorithm runs for each set of input parameters.

The results presented in Fig. 7.8 confirm our conviction that the Gaussian mutation $\alpha = 2$ is not the best choice in the general case. For distributions defined by the stability index $\alpha < 2$ and some scale parameter σ^*, the mean number of iterations needed to localize the neighbourhood of the global extremum, in the case of the benchmark functions considered, is significantly lower than in the case of algorithms with mutation based on the Gaussian distribution. Moreover, another general observation is worth noting: there exists, for each stability index α, some interval of scale parameters which guarantees high effectiveness of the $\mathrm{ESSS}_{S,\alpha}$ algorithm. For σ outside this interval, the effectiveness rapidly decreases. Because there is no possibility of direct transfer of methods of scale parameter adaptation created for the

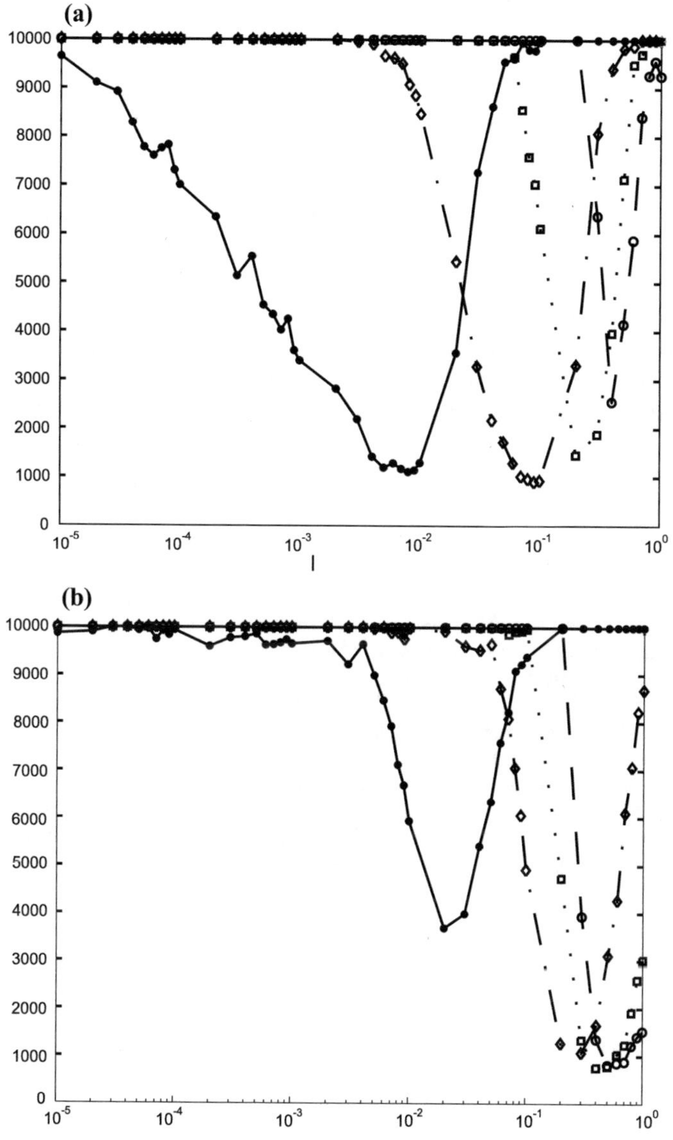

Fig. 7.8 Mean number of iterations needed for global extremum localization for Ackley's (**a**) and Rastringin's (**b**) functions versus the scale parameter σ: $\mathrm{ESSS}_{S,2}$ (circles), $\mathrm{ESSS}_{S,1.5}$ (quoters), $\mathrm{ESSS}_{S,1.0}$ (diamonds), $\mathrm{ESSS}_{S,0.5}$ (stars)

Gaussian distribution to distributions of lower values of the stability index, as well as by analyzing the results of our experiment, it can be suggested that, for a detailed optimization task, the stability index α can be adopted instead of the scale parameter σ. Perhaps such a process will be simpler to carry out.

7.5 Summary

Isotropic mutation with an α-stable generator allows minimization of the influence of the dead surrounding on the precision of local extremum localization. However, the $S\alpha SU(\sigma)$ distribution is not stable in general; it can be useful for solving optimization problems at least with the same effectiveness as in the case of the $IS\alpha S(\sigma)$ distribution. Especially, application of the $S\alpha SU(\sigma)$ distribution with low α to the mutation operator in evolutionary algorithms with soft selection, which guarantees enough offspring of the well-fitted parent element, seems to be very interesting. In this case, convergence to the local extremum with satisfying precision as well as the exploration of a relatively large area (due to frequent macro-mutations) is quite good. It is also worth noting that the efficiency of the evolutionary process is robust to a non-optimal choice of parameters which control the mutation process.

References

Karcz-Dulęba, I. (2001). Dynamics of infinite populations envolving in a landscape of uni- and bimodal fitness functions. *IEEE Transactions on Evolutionary Computation*, *5*(4), 398–409.

Karcz-Dulęba, I. (2004). Time to convergence of evolution in the space of population states. *International Journal of Applied Mathematics and Computer Science*, *14*(3), 279–287.

Kim, N. G., Won, J. M., Lee, J. S., Kim, S. W. (2002). Local convergence rate of evolutionary algorithm with combined mutation operator. *2002 Congress on Evolutionary Computation* (pp. 261–266). New York: IEEE Press.

Obuchowicz, A., & Prętki, P. (2005). Isotropic symmetric α-stable mutations for evolutionary algorithms. *IEEE Congress on Evolutionary Computation, Edinbourgh, UK* (pp. 404–410).

Rudolph, G. (1997). Local convergence rates of simple evolutionary algorithms with Cauchy mutations. *IEEE Transactions on Evolutionary Computation*, *1*(4), 249–258.

Chapter 8
Mutation Based on Directional Distributions

8.1 Motivation

Distributions of a spherical symmetry are probably the most frequently used distributions in mutation operators of evolutionary algorithms (Arabas 2001; Beyer and Schwefel 2002; Galar 1989; Törn and Zilinskas 1989). They, no doubt, possess both advantages and disadvantages. The main advantage is certainly a small number of parameters which explicitly define them, so configuration or adaptation procedures are based on clear and simple heuristics which model the current population surrounding. However, the small number of parameters is a limitation of probabilistic searching models that would be well-fitted to the currently solved problems because such models cannot include correlations between decision variables. The only physical parameter which can be reflected is the correlation between the fitness value of a sample and its Euclidean distance to a base point. If an adaptation landscape is characterized by more complicated and strong correlations between decision variables, then stochastic optimization with the isotropic exploration distribution necessarily implies a long time of the optimization process.

The elliptical normal distribution $\mathbf{N}(\boldsymbol{\mu}, \mathbb{C})$ (Beyer and Arnold 2003; Beyer and Schwefel 2002; Hansen and Ostermeyer 2001) is a technique which tries to overcome the above problem. This distribution allows modelling linear relations between decision variables. Thus, procedures which use the above distribution are usually enriched by an adaptation mechanism which has to fit the searching model to the population surrounding. But the point reflection in a base point, connected with the $\mathbf{N}(\boldsymbol{\mu}, \mathbb{C})$ distribution, results in the same probability of offspring generation in the desired direction and in the opposite, undesired one. Let us consider this problem in detail.

Despite the effectiveness of different heuristics, tuning the covariance matrix $\mathbb{C}$ of the $\mathbf{N}(\boldsymbol{\mu}, \mathbb{C})$ distribution leads to nearing the directions of the longest axis of the n-dimensional ellipsoid and of the fastest decline of the fitness function value. Figure 8.1 presents an idealised illustration of the above fact for a 2D spherical fitness function $f_{sph}(\boldsymbol{x})$ (B.1) and the approximation of the global minimum $\boldsymbol{x}_k = [2, 2]^T$.

A. Obuchowicz, *Stable Mutations for Evolutionary Algorithms*,
Studies in Computational Intelligence 797,
https://doi.org/10.1007/978-3-030-01548-0_8

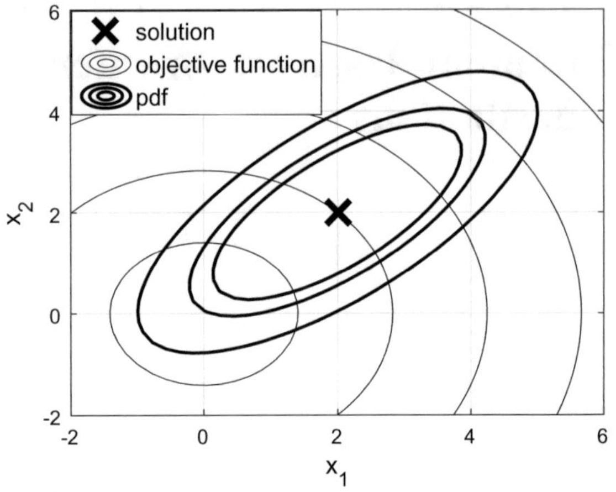

Fig. 8.1 Contour plot of the objective sphere function and the probability density function of the normal distribution, whose covariance matrix increases the probability of successful mutation of the point $\boldsymbol{x}_k = [2, 2]^T$

Analyzing the contour plots of the objective function and the probability density function of the normal distribution (Fig. 8.1), it is easy to see the most important disadvantage of the point reflection of the distribution: the probability of offspring location in the most attractive direction is the same as in the case of the opposite direction. As a result, over a half of mutations are not successful. This fact, undoubtedly, causes a significant decrease in the effectiveness of stochastic optimization algorithms, which is worsened with the searching space dimension increasing.

The increase of the searching space dimension is connected with a negative phenomenon for isotropic mutation. It is the so-called *course of dimensionality*. Let us look into it. In accordance with Theorem 3.17, each random vector $\boldsymbol{X}$ of a distribution of a spherical symmetry can be presented as a stochastic decomposition:

$$\boldsymbol{X} = r\boldsymbol{u}^{(n)}, \tag{8.1}$$

where $\boldsymbol{u}^{(n)}$ is a random vector uniformly distributed on the surface of the n-dimensional unit sphere and $r > 0$ is a random positive variable independent of $\boldsymbol{u}^{(n)}$. The relation between r and $\boldsymbol{X}$ is a bijection, thus the set of all isotropic distributions and that of all positive distributions are of the same cardinality (Fang et al. 1990). One can also point out that the vector $\boldsymbol{u}^{(n)}$ has the largest possible random entropy over all distributions defined on the unit sphere. It means that this distribution gives the largest amount of information about the space on which it is defined. This property is usually cited as an explanation of the application of the normal distribution in stochastic optimization algorithms in $\mathbb{R}^n$. At a first glance, it seems to be desired. In practice, it causes a rapid decrease in a stochastic searching process with the space dimension increasing. If we assume that there are only a few possible directions which lead to solutions of some desired fitness, then we must remember that the largest entropy describes the largest uncertainty of selection of these directions. It can be assumed, without loss of generality, that there exists one most beneficial dimension in terms of exploitation. In the case of differentiable problems, this beneficial dimension is opposite to the gradient of an objective function. In the case of non-differentiable functions, it is a direction taken from the neighbourhood of the base point, which guarantees the greatest improvement of solution quality. In most cases, each deviation from this direction causes significant deterioration of solution quality. The above intuitive discussion can be formalized as the theorem presented below (Prętki and Obuchowicz 2006).

Theorem 8.1 *Let X be an n-dimensional random vector of a distribution with a spherical symmetry. Let us assume that an alternative solution is generated as an additive disturbance using the vector $\boldsymbol{X}$, i.e., $\boldsymbol{x}_{t+1} = \boldsymbol{x}_t + \boldsymbol{X}$. Moreover, let $\boldsymbol{\mu}_t$ describe the most beneficial, in the sense of fitness improvement, direction of the $\boldsymbol{x}_t$ mutation. Thus, the limit of the probability that the solution $\boldsymbol{x}_{t+1}$ is localized in a hyperplane perpendicular to $\boldsymbol{\mu}_t$ is equal to 1 as n approaches infinity.*

Proof Let $\boldsymbol{\mu} \in \mathbb{R}^n$ be normalized: $\|\boldsymbol{\mu}\| = 1$, and let $\boldsymbol{u}^{(n)}$ denote a random vector uniformly distributed on the unit sphere surface. Then, the distribution of the random variable $\nu = \boldsymbol{\mu}^T \boldsymbol{u}^{(n)}$ is described by the following probability density function (Fang et al. 1990):

$$p_n(\nu) = \frac{1}{\beta(\frac{n-1}{2}, \frac{1}{2})} (1 - \nu^2)^{\frac{n-3}{2}}, \tag{8.2}$$

where $\beta(\cdot, \cdot)$ denotes the Beta function (Rudin 1976). Let us consider the limit of the function sequence $\{p_n\}$ for $\nu \in (-1, 0) \cup (0, 1)$:

$$\lim_{n\to\infty} p_n(\nu) = \pi^{-1/2} \lim_{n\to\infty} \frac{\Gamma(\frac{n}{2})}{\Gamma(\frac{n-1}{2})} (1 - \nu^2)^{\frac{n-3}{2}}. \tag{8.3}$$

Substituting $k = 2(n + 1)$ and $b = (1 - \nu^2) \in (0, 1)$, (8.3) has the form

$$\lim_{n\to\infty} p_n(t) = \lim_{k\to\infty} \frac{\Gamma(k + 1)}{\Gamma(k + 1/2)} b^{k-1/2}. \tag{8.4}$$

Next, let us note that the following inequality is true:

$$a_k = 0 \le \frac{\Gamma(k+1)}{\Gamma(k+1/2)} b^{k-1/2} < (k + 1/2)^{1/2} b^{k-1/2} \le (k + 1/2) b^{k-1/2} = c_k.$$

The limit of the sequence c_k can be calculated using de l'Hospital's rule:

$$\lim_{k\to\infty} c_k = \lim_{k\to\infty} \frac{k + 1/2}{b^{1/2-k}} = \lim_{k\to\infty} \frac{1/2}{-\ln(b) b^{1/2-k}} = -\frac{b^{-1/2}}{2\ln(b)} \lim_{k\to\infty} b^k = 0.$$

Both sequences a_k, c_k converge to zero, and from the squeeze theorem (Rudin 1976) it follows that the limit of the sequence d_k is also equal to zero. Thus, the probability density function (8.2) converges to zero for each $\nu \in [-1, 0) \cup (0, 1]$ and approaches infinity for $\nu = 0$. A consequence of this fact is concentration of the whole probability

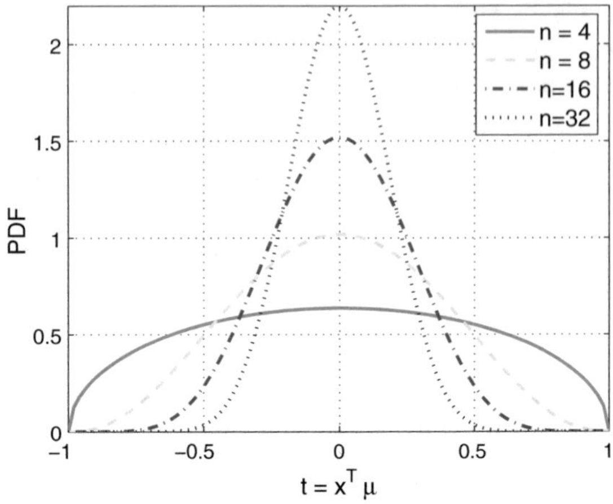

Fig. 8.2 Probability density function of the random variable $\nu = \boldsymbol{Z}^T \boldsymbol{\mu}$ for the isotropic exploration vector $\boldsymbol{Z}$ and the most beneficial mutation direction $\boldsymbol{\mu}$

mass in the point $\nu = 0$. If the vector $\boldsymbol{\mu}$ describes the most beneficial direction in the searching space, then we prove our theorem. □

Figure 8.2 illustrates Theorem 8.1.

Theorem 8.1 allows us to explain many phenomena observable in the results previously mentioned in this book, especially in Chap. 5. For example, analyzing the local convergence of the $(1+1)\mathrm{ES}_{I,\alpha}$ strategy (Sect. 5.3.1), the increase of the mean number of iterations needed for solution localization can be seen (Fig. 5.5). This phenomenon is difficult to explain without Theorem 8.1, all the more so as the Euclidian distance between the starting point and the global solution is constant and independent of the searching space dimension n. Similarly, analyzing the saddle crossing ability of the $(1, 2)\mathrm{ES}_{I,\alpha}$ strategy (Sect. 5.3.2), the course of dimensionality described in Theorem 8.1 can be observed in the narrowing of the range of scale parameter values, which guarantee successful experiment processing (Fig. 5.7). Due to the elitist selection character in the algorithm considered, the only mechanism which causes saddle crossing is macro-mutations. If the space dimension increases, the proportion of the solid angle of successful mutation and the whole surface of the unit sphere decreases.

An attempt to diminish the above-described phenomena is proposed by Obuchowicz (2003a). The classical version of the ESSS algorithm (Table 2.3) (Galar 1989) is enriched by the *forced direction of mutation* (FDM) mechanism. The idea is based on the rule stating that, if natural conditions of the adaptation landscape prefer some direction of the population drift, then this direction is supported not only by the selection but also by the mutation operator, using the following schema:

$$\boldsymbol{x}_{k+1}^{(i)} = \boldsymbol{x}_k^{(i)} + N(\boldsymbol{\mu}, \sigma \boldsymbol{I}_n), \tag{8.5}$$

$$\boldsymbol{\mu} = c\,\sigma \frac{\langle x_k \rangle - \langle x_{k-1} \rangle}{\| \langle x_k \rangle - \langle x_{k-1} \rangle \|}, \tag{8.6}$$

$$\langle x_k \rangle = \frac{1}{\eta} \sum_{i=1}^{\eta} \boldsymbol{x}_k^{(i)}, \tag{8.7}$$

where $\{\boldsymbol{x}_k^{(i)} \mid i = 1, 2, \ldots, \eta\}$ is the set of elements in the k-th iteration, while σ and c are arbitrarily chosen parameters. As follows from (8.5)–(8.7), the preferred direction of mutation is determined using the selection operator, which modifies the centers of gravity of subsequent populations by elimination of worse elements. This algorithm, previously created for non-stationary optimization problems, proves its big efficacy in problems of dynamic neural network parametric estimation (Obuchowicz 2003b).

The above idea is expanded in this chapter with a somewhat different approach: directional distributions are applied (Mardia and Jupp 2000).

8.2 Directional Distributions

Exploration directional distributions are based on a similar stochastic decomposition like in the case of isotropic distributions. The difference is that the random vector $\boldsymbol{u}^{(n)}$ of a uniform distribution on the surface of the n-dimensional unit sphere is substituted by the random vector $\boldsymbol{d}^{(n)}$ of a non-uniform distribution on this sphere surface (Prętki and Obuchowicz 2006):

$$\boldsymbol{X} = r\, \boldsymbol{d}^{(n)}. \tag{8.8}$$

8.2.1 Von Mises–Fisher Distributions

Between many probabilistic models of the above type (Mardia and Jupp 2000), von Mises–Fisher distributions $MF(\boldsymbol{\mu}, \kappa)$ of a rotational symmetry (Mardia 1999) seem to be most important in stochastic optimization processes. They are clearly defined using two parameters: the preferred direction $\boldsymbol{\mu}$, $\|\boldsymbol{\mu}\| = 1$ and the dispersion $\kappa \geq 0$. The probability density function of the n-dimensional distribution $MF(\boldsymbol{\mu}, \kappa)$ has the form (Dhillon and Sra 2003)

$$p_n(\boldsymbol{s}; \kappa, \boldsymbol{\mu}) = \frac{\kappa^{n/2-1}}{(2\pi)^{n/2} I_{\frac{n}{2}-1}(\kappa)} \exp(\kappa \boldsymbol{\mu}^T \boldsymbol{s}), \tag{8.9}$$

where $\boldsymbol{s} \in S_n$ is a point on the surface of the unit sphere. The direction $\boldsymbol{\mu}$ is preferred by the distribution $MF(\boldsymbol{\mu}, \kappa)$. In the case of application of this distribution to stochastic optimization algorithms, the direction $\boldsymbol{\mu}$ can be chosen using the same techniques as for establishing the longest axe of the elliptical normal distribution $\mathbf{N}(\boldsymbol{\mu}, \mathbb{C})$. The frequency of $\boldsymbol{\mu}$ direction sampling is controlled by the dispersion parameter κ and can be arbitrarily chosen. So, κ decides how quickly the probability of alternative solution localization in the $\hat{\boldsymbol{\mu}}$ direction decreases with the angle between $\boldsymbol{\mu}$ and $\hat{\boldsymbol{\mu}}$

increasing. If the value of κ is properly chosen, then the problem described in the comments to Fig. 8.1 can be successfully solved.

In order to verify the applicability of von Mises–Fisher distributions to the mutation operation in evolutionary algorithms, let us consider a simulation experiment. We try to localize the extremum of the sphere function $f_{sph}(\boldsymbol{x})$ (B.1) using the simplest evolutionary strategy $(1+1)$ES. The classical version of this algorithm uses the $N(\boldsymbol{0}, \sigma \boldsymbol{I}_n)$ distribution in the mutation operator, and the random vector of this distributions can be stochastically decomposed as follows:

$$\boldsymbol{X} \stackrel{d}{=} \boldsymbol{u}^{(n)} \sigma \chi_n, \tag{8.10}$$

where χ_n is a random variable with n degrees of freedom. We create the mutation operator similarly as in (8.10), and the following vector is added to the base one:

$$\boldsymbol{X} \stackrel{d}{=} \boldsymbol{d}^{(n)} \sigma \chi_n, \tag{8.11}$$

where $\boldsymbol{d}^{(n)} \sim MF(\boldsymbol{\mu}, \kappa)$. The comparison of the convergence of two strategies $(1+1)$ES which have mutation operators based on the distributions (8.10) and (8.11) for different space dimensions n is the goal of our experiment.

Let us assume that an ideal adaptation heuristic, which perfectly chooses the most preferable mutation direction $\boldsymbol{\mu}^*$, is known. In our numerical calculations we utilise the differentiability of the sphere function $f_{sph}(\boldsymbol{x})$ (B.1), and this direction is opposite to the gradient

$$\boldsymbol{\mu}^* = -\nabla f_{sph}(\boldsymbol{x}_k). \tag{8.12}$$

Let us choose the following starting conditions for the strategies considered: the starting point $\boldsymbol{x}_0 = [10000, 0, 0, \ldots, 0]^T$, the mutation scale $\sigma = 0.01$, the maximal number of iterations $t_{\max} = 10000$, the dispersion of the directional distributions $MF(\boldsymbol{\mu}, \kappa)$ $\kappa = 3$. A typical optimization process of the evolutionary strategy is shown in Fig. 8.3a and b. Moreover, Fig. 8.3c and d present histograms of $\boldsymbol{X}^T \boldsymbol{\mu}^*$, where $\boldsymbol{X}$ is the distribution vector (isotropic and directional, respectively) and $\boldsymbol{\mu}^*$ is the direction which guarantees the best convergence of algorithms (8.12).

The results presented in Fig. 8.3 prove the advantage of directional distributions, especially in the case of low space dimensions. Unfortunately, this advantage degrades with n increasing. This phenomenon is illustrated by histograms (Fig. 8.3d. Retaining the same value of the dispersion κ, the number of vectors generated in remote directions from the preferred $\boldsymbol{\mu}^*$ rapidly increases with n increasing. The histogram for $n = 32$ is almost identical as that for isotropic mutation, and most mutated vectors are located in the surface perpendicular to the direction $\boldsymbol{\mu}^*$. This fact is surprising, particularly when the probability density function of the $MF(\boldsymbol{\mu}, \kappa)$ distributions achieves the highest value for the direction $\boldsymbol{\mu}^*$. So, having an ideal heuristic indicating the most preferable direction of mutation, the $MF(\boldsymbol{\mu}, \kappa)$ distributions do not allow full use of this valuable information.

The fact described above questions the validity of the von Mises–Fisher distributions for mutation operations in evolutionary algorithms, especially in the case of

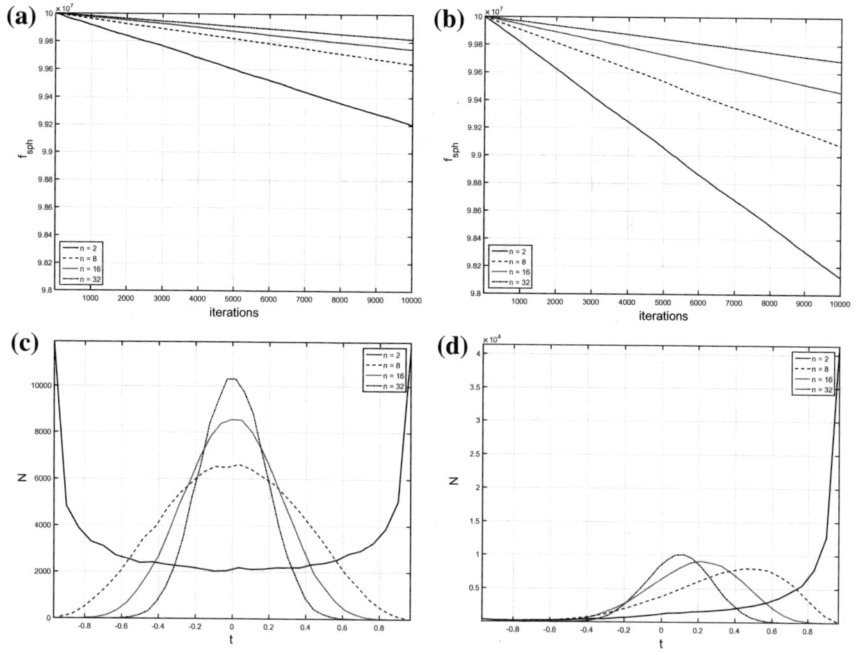

Fig. 8.3 Sphere function minimization process of (1 + 1)ES with isotropic (**a**) and directional (**b**) mutations. Figures (**c**) and (**d**) present histograms of the random variable $t = \boldsymbol{X}^T \boldsymbol{\mu}^*$ for the results shown in (**a**) and (**b**), respectively

high-dimensioned searching spaces. Let us try to find a reason for this. The probability density function of the random variable $t = \hat{\boldsymbol{\mu}}^T \boldsymbol{d}^{(n)}$, where $\boldsymbol{d}^{(n)} \sim MF(\boldsymbol{\mu}, \kappa)$ and $\hat{\boldsymbol{\mu}}$ is a some normalized vector $\|\hat{\boldsymbol{\mu}}\| = 1$, has the form (Mardia and Jupp 2000)

$$p_n(t; \kappa) = \frac{(\frac{\kappa}{2})^{n/2-1}}{\Gamma(\frac{n-1}{2})\Gamma(n/2)I_{\frac{n-1}{2}}(\kappa)} \exp(\kappa t)(1-t^2)^{\frac{n-3}{2}}, \tag{8.13}$$

where $\|\boldsymbol{\mu}\|_2 = 1$, $I_n(\cdot)$ is a modified Bessel function of the first order. The expectation value $E[t] = E[\hat{\boldsymbol{\mu}}^T \boldsymbol{d}^{(n)}]$ is decreasing with n increasing (Prętki and Obuchowicz 2006). Thus, despite the fact that the $MF(\boldsymbol{\mu}, \kappa)$ distribution prefers the direction $\boldsymbol{\mu}$, directions of the pseudo-random vectors generated by this distribution are further and further away from the direction $\boldsymbol{\mu}$ with the space dimension n increasing.

Thus, we undertake the attempt at construction of a distribution with a rotational symmetry with respect to any angle around a preferred direction $\boldsymbol{\theta}$. This distribution has any marginal distribution $t = \boldsymbol{\mu}^T \boldsymbol{X}$. Each random vector of this distribution can be described by the following stochastic decomposition (Mardia and Jupp 2000):

$$\boldsymbol{X} = t\boldsymbol{\theta} + \sqrt{1-t^2}\boldsymbol{\xi}, \tag{8.14}$$

where t is a random variable independent of rotations around $\boldsymbol{\theta}$, and $\boldsymbol{\xi}$ has a uniform distribution on the surface of the unit sphere $\partial S_1^{(n-2)}(\mathbf{0})$. Moreover, $\boldsymbol{\xi}$ and t are independent of each other. The decomposition (8.14) can be used to construct the distribution of a rotational symmetry with any marginal distribution. In order to do it, let us substitute $\boldsymbol{\theta} = [0, 0, \ldots, 1]^T \in \mathbb{R}^n$ in (8.14) and consider the orthogonal transformation $\boldsymbol{QX}$, such that $\boldsymbol{Q\theta} = \boldsymbol{\mu}$. It can be simply proved that the random vector obtained in this way has the marginal distribution $\boldsymbol{X}^T\boldsymbol{\theta} \overset{d}{=} t$. Choosing in an appropriate way the distribution of the random variable t, the amount of the probability mass focused around the expected direction $\boldsymbol{\mu}$ can be controlled in any way (Prętki and Obuchowicz 2006). Below, the application of the marginal distribution $t = 2X - 1$, where X is a random variable of the Beta distribution $\beta(a, b)$ (Shao 1999), is recommended. Thus, the probability density function of the variable t is described by (Prętki and Obuchowicz 2006)

$$p(t|a, b) = \frac{2^{1-a-b}}{\beta(a, b)}(1-t)^{b-1}(1+t)^{a-1}, \tag{8.15}$$

where a and b are parameters of the Beta distribution. The dispersion parameter, which is used in von Misses–Fisher distributions, can also be used in the proposed distributions by application of suitable values of parameters of the distribution (8.15), i.e., $a = \frac{n-1}{2}$ and $b = \kappa\frac{n-1}{2}$. The expectation value of the random variable $T = \boldsymbol{\theta}^T\boldsymbol{X}$ and its variance are equal to

$$\mathrm{E}[T] = \frac{1-\kappa}{1+\kappa}, \qquad \mathrm{Var}(T) = \frac{8\kappa}{n(1+k)^2(1-\kappa^2)}. \tag{8.16}$$

The expectation value (8.16) of the marginal distribution $\boldsymbol{\theta}^T\boldsymbol{X}$ is independent of the space dimension n. Meanwhile, the space dimension occurs in the definition of variance. It causes, in the case of the directional distributions considered, the increase of the number of pseudo-random realisations around the preferred direction with the space dimension increasing. This effect especially clearly manifests itself in Fig. 8.4. Thanks to the assumed parametrization, the dispersion parameter κ allows us to obtain a uniform distribution over the surface of the unit sphere for $\kappa = 1$, and the distribution generated for the vector $\boldsymbol{\mu}$ for $\kappa = 0$ (Prętki and Obuchowicz 2006). The main advantage of the proposed directional distribution is a simple algorithm of generation of pseudo-random vectors (Table 8.1). Figure 8.5 presents realizations of directional distributions in accordance with the algorithm shown in Table 8.1.

Another question connected with application of the directional distribution to the mutation operator is selection of the mutation variable $r > 0$ (8.11). It is worth noting that, in the case of the stochastic decomposition of the directional distribution $\boldsymbol{Z} \sim M(\boldsymbol{\mu}, \kappa)$, the distribution of the norm is $\|\boldsymbol{Z}\| \overset{d}{=} r$. This means that the burdensome dead surrounding effect can be eliminated if the mutation variable $r > 0$ is independent of the searching space dimension. The experimental part of this chapter concerns application of stable distributions as distributions of the mutation variable r.

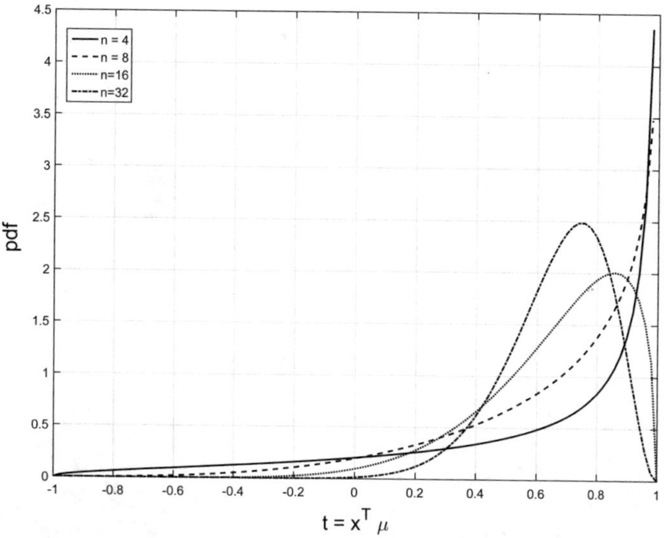

Fig. 8.4 Probability density function of the marginal distribution $t = \boldsymbol{X}\boldsymbol{\mu}$ for the proposed directional distribution $\boldsymbol{X} \sim M(\boldsymbol{\mu}, \kappa)$

Table 8.1 Algorithm of generation of pseudo-random vectors of the directional distribution $M(\boldsymbol{\mu}, \kappa)$

Input

$\mu \in \mathbb{R}^n$: preferred direction

$\kappa \in (0, 1]$: dispersion parameter

Output

Y: pseudo-random vector of the $M(\mu, \kappa)$ distribution

Algorithm

$t = 2\beta(\frac{n-1}{2}, \frac{\kappa(n-1)}{2}) - 1$, $\beta(a,b)$ random variable of the Beta distribution

$X \leftarrow \mathbf{N}(0, I_{n-1})$

$Z \leftarrow X/\|X\|_2$

$Y \leftarrow [\sqrt{1-t^2}Z^T, t]^T$

$Y \leftarrow [I_n - 2vv^T]Y$ where $v = \frac{[0,0,\ldots,1]^T - \mu}{\|[0,0,\ldots,1]^T - \mu\|_2}$

8.3 Simulation Experiments

Evolutionary search with proportional selection (ESSS) (Table 2.3) and with tournament selection (ESTS) are chosen as the basic algorithms for simulation experiments verifying the influence of the directional mutation $M(\boldsymbol{\mu}, \kappa)$ on the effectiveness of the optimization process. The only difference between standard versions of both algorithms and those proposed in this experiment is application of the directional distribution $M(\boldsymbol{\mu}, \kappa)$ to the mutation operator instead of the Gaussian distribution.

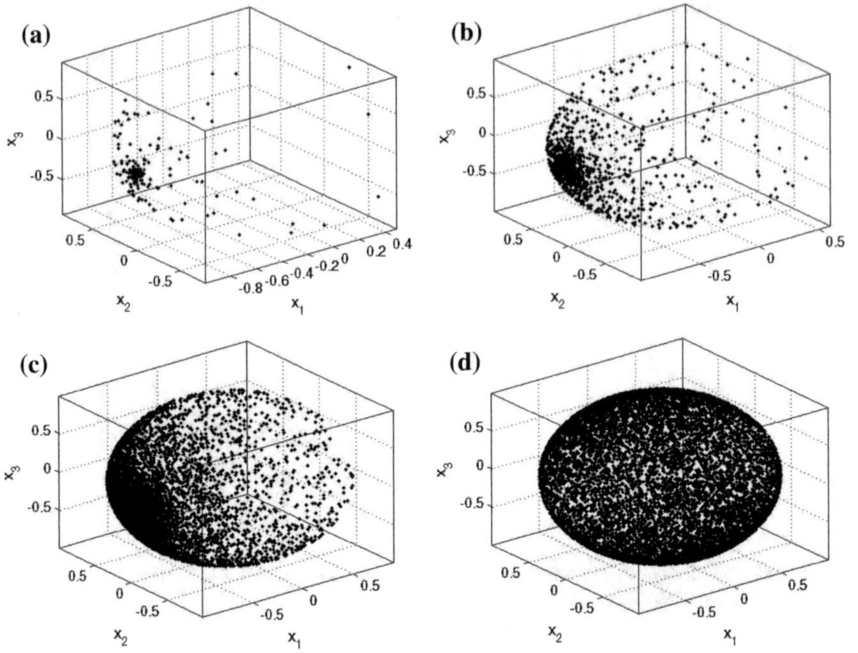

Fig. 8.5 Realizations of 10000 pseudo-random vectors of the directional distribution $M(\boldsymbol{\mu}, \kappa)$ for $\boldsymbol{\mu} = [-1, 0, 0]^T$ and $\kappa = 0.001$ (**a**), $\kappa = 0.01$ (**b**), $\kappa = 0.1$ (**c**), $\kappa = 1$ (**d**)

The final versions of the algorithms described by ESSS$_\alpha$-DM and ESTS$_\alpha$-DM are presented in Table 8.2.

The basic problem which is connected with introduction of directional mutation is the way of selecting the preferred direction $\boldsymbol{\mu}$. Assuming that, generally, there is no possibility to calculate the gradient vector of the objective function $f(\boldsymbol{x})$ in set points or its numerical approximation, three heuristics of the direction $\boldsymbol{\mu}$ selection are examined:

- **Heuristic no. 1**:

$$\boldsymbol{\mu}^t = -\frac{z_t}{\|z_t\|_2}, \quad \text{where} \quad z_t = [\boldsymbol{P}_t^T \boldsymbol{P}_t]^{-1} \boldsymbol{P}_t^T \boldsymbol{F}_t, \tag{8.17}$$

where $\boldsymbol{P}_t \in R^{\eta \times n}$ describes the matrix with rows containing phenotypes of population elements in iteration t and the corresponding fitness vector $\boldsymbol{F}_t \in R^\eta$,

$$\boldsymbol{P}_t = [\boldsymbol{x}_1^t, \boldsymbol{x}_2^t, \ldots, \boldsymbol{x}_\eta^t]^T, \qquad \boldsymbol{F}_t = [f(\boldsymbol{x}_1^t), f(\boldsymbol{x}_2^t), \ldots, f(\boldsymbol{x}_\eta^t)]^T.$$

Table 8.2 Evolutionary search algorithm with soft proportional selection ESSS$_\alpha$-DM (tournament selection ESTS$_\alpha$-DM) and forced direction of mutation

Input

η: population size;

$T_{\max}$: maximal number of iterations;

σ, α, κ: scale parameter, stability index, dispersion;

$f : \mathbb{R}^n \to \mathbb{R}$: objective function;

x_0^0: starting point.

Algorithm

1. Initiation

$$P(0) = \left(x_1^0, x_2^0, \ldots, x_\eta^0\right), \; x_k^0 = x_0^0 + Z, \quad \text{where } \; Z \sim \mathbf{N}(0, I_n), \; k = 1, 2, \ldots, \eta.$$

2. Repeat

(a) Evaluation

$$F\big(P(t)\big) = \left(q_1^t, q_2^t, \ldots, q_\eta^t\right), \; \text{ where } \; q_k^t = f\left(x_k^t\right), \quad k = 1, 2, \ldots, \eta.$$

(b) Selection (proportional or tournament)

$$P(t) \longrightarrow P(t)' = \left(x_{h_1}^t, x_{h_2}^t, \ldots, x_{h_\eta}^t\right).$$

(c) Estimation of the mutation direction using the heuristics (8.17)–(8.19)

$$\mu(t) \longleftarrow H\big(P'(t), P'(t-1)\big).$$

(d) Mutation

$$P(t)' \longrightarrow P(t+1);$$

$$x_k^{t+1} = x_{h_k}^t + \chi_{\alpha,\sigma} d^{(n)}, \, d^{(n)} \sim M(\mu(t), \kappa),$$

$$\chi_{\alpha,\sigma} \sim |S_\alpha S(\sigma)| \, k = 1, 2, \ldots, \eta.$$

Until $t > t_{\max}$.

- **Heuristic no. 2**:

$$\boldsymbol{\mu}^t = -\frac{z_t}{\|z_t\|_2}, \quad \text{where } \; z_t = \sum_{k=1}^{\eta} \frac{f(\boldsymbol{x}_k^{t-1}) - f(\boldsymbol{x}_k^t)}{f(\boldsymbol{x}_k^{t-1})} \frac{\boldsymbol{x}_k^t - \boldsymbol{x}_k^{t-1}}{\|\boldsymbol{x}_k^t - \boldsymbol{x}_k^{t-1}\|_2}. \tag{8.18}$$

- **Heuristic no. 3**:

$$\boldsymbol{\mu}^t = \frac{\langle \boldsymbol{x}^t \rangle - \langle \boldsymbol{x}^{t-1} \rangle}{\|\langle \boldsymbol{x}^t \rangle - \langle \boldsymbol{x}^{t-1} \rangle\|}, \quad \text{where } \; \langle \boldsymbol{x}^t \rangle = \frac{1}{\eta} \sum_{k=1}^{\eta} \boldsymbol{x}_k^t. \tag{8.19}$$

The heuristic (8.17) concerns elements of the population obtained by selection as a set of approximation nodes, which are used to search for the best fitted (in the mean square sense) multidimensional linear function. The direction $\boldsymbol{\mu}$ is identified with the direction of the steepest descent pointed out by linear approximation of the population distribution. One of the disadvantages of this method is the necessity of inverse matrix calculation, so a suitable size of the population should be guaranteed in order to perform this operation. The heuristic (8.18) is proposed by Salomon (1998) for algorithms known as *evolutionary gradient search*. The main advantage of this heuristic is cumulation of experiences from each sample of random solution improving. The heuristic (8.19) is taken from the above-mentioned evolutionary search with soft selection and forced direction of mutation algorithm proposed by Obuchowicz (2003b).

8.3.1 Conducting the Experiment

The aim of the experiment is the comparison of evolutionary algorithms (Table 8.2) based on directional mutations and the heuristics (8.17)–(8.19) assisting them. The research is focused on the most popular testing functions: the sphere function $f_{sph}(\boldsymbol{x})$ (B.1), the generalized Rastringin function $f_R(\boldsymbol{x})$ (B.4), the Ackley function $f_A(\boldsymbol{x})$ (B.3), and the ellipsoidal function $f_e(\boldsymbol{x})$ (B.6), .

Algorithms of the classes ESSS$_\alpha$-DM and ESTS$_\alpha$-DM are explicitly defined by three parameters: α, σ and κ. In order to perform a detailed analysis of correlations between these parameters and the heuristics applied, a series of experiments are done for all possible sets of the following values:

$$\alpha = \{2.0, 1.75, 1.5, 1.25, 1.0, 0.75, 0.5\},$$
$$\sigma = \{0.01, 0.05, 0.1, 0.2, 0.5, 1, 2, 5, 10\},$$
$$\kappa = \{0.01, 0.1, 0.25, 0.5, 0.75, 1\}.$$

Each parameter configuration $\{\alpha, \sigma, \kappa\}$ and one of the heuristics considered is evaluated based on 100 independent evolutionary algorithm runs. Initial populations for each algorithm version are created by normal mutations $\mathbf{N}(\mathbf{0}, \boldsymbol{I}_n)$ of the same point $\boldsymbol{x}_0$. Soft selection and a lack of the mutation adaptation mechanism are the reason why these algorithms are not convergent to global solutions. They strive to the selection–mutation equilibrium at some distance from the extremum (Karcz-Dulęba 2004). This is why specific stop criteria are applied. First of all, the maximum of the objective function evaluation is chosen: $t_{\max} = 10000$ for f_{sph} and f_e and $t_{\max} = 100000$ for f_R and f_A. Moreover, algorithms are stopped when the best solution does not improve in a given interval of time: 200 iterations for unimodal functions and 500 iterations for multi-modal functions. The population size $\eta = 50$ is chosen to avoid problems with inverse matrix operations, which are used in Heuristic no. 1.

8.3.2 Results

An important preliminary issue is selection of the most advantageous set of the analyzed algorithms' parameters $\{\alpha, \sigma, \kappa\}$ for each optimization problem. The minimization of the median of the values of the objective function generated by each algorithm ESSS$_\alpha$-DM or ESTS$_\alpha$-DM is chosen as a criterion of comparison of each set $\{\alpha, \sigma, \kappa\}$ and each heuristic. The optimal sets of $\{\alpha, \sigma, \kappa\}$ for 2D objective functions and algorithms of the class ESTS$_\alpha$-DM are presented in Table 8.3. Tables 8.6 and 8.9 show results for 30D versions of the tested problems and the algorithms ESSS$_\alpha$-DM and ESTS$_\alpha$-DM, respectively.

The first fact which draws our attention when analyzing optimal parameters obtained for 2D testing problems (Table 8.3) is a relatively high value of κ. Application of a numerous population ($\eta = 50$) in comparison with the space dimension makes the heuristic selection question not so important. The macro-mutation mechanism guaranteed by stability distributions with a low stability index α has crucial impact on the ESTS$_\alpha$-DM algorithm's effectiveness. This observation is confirmed by the results given in Table 8.4, where positive correlation factors between the stability index α and medians of the objective functions f_{sph}, f_R and f_A clearly show the advantage of distributions with heavy tails. It is worth noting that there is no clear correlation between the dispersion κ and the obtained results (Table 8.5). Taking into account a large population in comparison with the space dimension, it can be

Table 8.3 Optimal parameters $\alpha^*, \kappa^*, \sigma^*$ obtained for the algorithm ESTS$_\alpha$-DM and different heuristics of determining the mutation direction. $MNOFE$: median of the number of objective function evaluation, $MOFV$: median of the objective function value. Searching space dimension $n = 2$

$f(\cdot)$	α^*	κ^*	σ^*	MOFV	MNOFE
ESTS$_\alpha$-DM + Heuristic no. 1					
f_{sph}	0.5	1	0.01	$2.4487e-007$	2495
f_e	1	0.75	0.1	0.24094	328.5
f_R	1.5	0.1	0.5	1.0544	373.5
f_A	0.5	0.25	0.1	0.033006	917.5
ESTS$_\alpha$-DM + Heuristic no. 2					
f_{sph}	0.5	1	0.01	$3.0264e-007$	2455
f_e	1.99	0.75	0.1	0.1708	302
f_R	0.5	0.5	0.1	1.0176	567.5
f_A	0.5	0.75	0.1	0.067061	692.5
ESTS$_\alpha$-DM + Heuristic no. 3					
f_{sph}	0.5	1	0.01	$2.0941e-007$	2517
f_e	1.99	0.1	0.01	$1.3824e-005$	2355
f_R	0.5	0.75	0.1	1.0094	541
f_A	0.5	0.5	1	0.37435	533.5

Table 8.4 Linear correlation factors for independent variables α, σ and three explanatory variables $Me[H_1]$, $Me[H_2]$, $Me[H_3]$. Searching space dimension $n = 2$ and the ESTS$_\alpha$-DM algorithm

Objective function f_{sph}

Variables	H_1, 0.01	H_2, 0.01	H_3, 0.01	H_1, 1.00	H_2, 1.00	H_3, 1.00
α	0.31	0.31	0.30	0.33	0.34	0.33
σ	–0.23	–0.21	–0.23	–0.25	–0.24	–0.25

Objective function f_e

Variables	H_1, 0.01	H_2, 0.01	H_3, 0.01	H_1, 1.00	H_2, 1.00	H_3, 1.00
α	0.12	0.09	0.14	0.08	0.09	0.08
σ	0.95	0.87	0.93	0.95	0.96	0.92

Objective function f_R

Variables	H_1, 0.01	H_2, 0.01	H_3, 0.01	H_1, 1.00	H_2, 1.00	H_3, 1.00
α	0.26	0.33	0.33	0.37	0.38	0.38
σ	–0.26	–0.25	–0.28	–0.21	–0.21	–0.21

Objective function f_A

Variables	H_1, 0.01	H_2, 0.01	H_3, 0.01	H_1, 1.00	H_2, 1.00	H_3, 1.00
α	0.50	0.30	0.27	0.45	0.38	0.39
σ	–0.51	0.73	0.46	–0.64	–0.68	–0.67

Table 8.5 Linear correlation factors for independent variables κ, σ and three explanatory variables $Me[H_1]$, $Me[H_2]$, $Me[H_3]$: medians of the objective function values are obtained over 100 independent ESTS$_\alpha$-DM algorithm runs with the heuristics H_1, H_2, H_3. Searching space dimension $n = 2$

Objective function f_{sph}

Variables	H_1, 1.99	H_2, 1.99	H_3, 1.99	H_1, 0.50	H_2, 0.50	H_3, 0.50
κ	0.03	0.05	0.04	–0.33	–0.25	–0.28
σ	–0.39	–0.39	–0.39	–0.11	–0.13	–0.14

Objective function f_e

Variables	H_1, 1.99	H_2, 1.99	H_3, 1.99	H_1, 0.50	H_2, 0.50	H_3, 0.50
κ	–0.05	–0.09	–0.14	0.01	–0.04	–0.06
σ	0.95	0.90	0.93	0.94	0.90	0.92

Objective function f_R

Variables	H_1, 1.99	H_2, 1.99	H_3, 1.99	H_1, 0.50	H_2, 0.50	H_3, 0.50
κ	0.03	0.04	0.01	–0.20	0.04	–0.09
σ	–0.42	–0.41	–0.41	0.66	0.76	0.59

Objective function f_A

Variables	H_1, 1.99	H_2, 1.99	H_3, 1.99	H_1, 0.50	H_2, 0.50	H_3, 0.50
κ	0.14	–0.16	–0.14	0.20	–0.58	–0.55
σ	–0.88	–0.71	–0.76	–0.21	–0.13	–0.30

Table 8.6 Optimal parameters α^*, κ^*, σ^* obtained for the ESSS$_\alpha$-DM algorithm with directional mutation and different heuristics of determining the mutation direction. $MNOFE$: median of the number of objective function evaluation, $MOFV$: median of the objective function value. Searching space dimension $n = 30$

$f(\cdot)$	α^*	κ^*	σ^*	MOFV	MNOFE
ESSS$_\alpha$-DM + Heuristic no. 1					
f_{sph}	2.0	0.1	5	505.4189	6794.5
f_e	2.0	0.25	0.1	3950.4333	2572.5
f_R	2.0	0.1	2	244.6408	2782.5
f_A	2.0	0.1	10	11.6101	3184.5
ESSS$_\alpha$-DM + Heuristic no. 2					
$f(\cdot)$	α^*	κ^*	σ^*	MOFV	MNOFE
f_{sph}	2.0	0.1	5	7596.2018	6924.5
f_e	2.0	0.1	1	18778.1374	341
f_R	2.0	0.5	2	406.4481	2832
f_A	2.0	0.5	0.01	19.9727	19607
ESSS$_\alpha$-DM + Heuristic no. 3					
$f(\cdot)$	α^*	κ^*	σ^*	MOFV	MNOFE
f_{sph}	2.0	0.1	5	31618.8829	6651
f_e	2.0	0.25	1	24163.9718	305
f_R	2.0	0.75	2	416.7339	3182
f_A	2.0	0.75	0.01	19.9731	26814

supposed that there are at least a couple of mutations in the proper direction regardless of its selection. In such a situation, the advantage of macro-mutations, which cause a faster shift to a global extremum neighborhood, is not surprising. Results obtained for the ill-conditioned problem f_e are a derogation from the above rule. In this case, the role of the convergence of the probability of success $P_s \to 0.5$ for $\sigma \to 0$ is dominating. The existence of a large population makes application of a mutation operator, which gives small, but often successful, shifts in the searching space, the most effective solution. One can find a confirmation of this observation in the unusually strong correlation between the scale parameter and results obtained for the function f_e (Tables 8.4 and 8.5).

Optimal sets of parameters $\{\alpha, \sigma, \kappa\}$ for the 30D searching space and the ESSS$_\alpha$-DM algorithm presented in Table 8.6 are totally dominated by the normal distribution. The cause can be found in the selection pressure decrease connected with macro-mutations. Taking into account the results presented in Table 8.7, it can be seen that the importance of the dispersion κ increases with the searching space size. Especially strong correlations are noted for distributions with low stable indices α. The most effective heuristic is heuristic no. 1. In the case of the multi-modal functions f_R and f_A, algorithms converge prematurely independently of the chosen heuristics. But a large number of iterations performed before stop criterion fulfillment allows calculating reliable statistical relationships between particular parameters. Analyzing the correlation factors shown (Table 8.8), it can be observed that macro-mutations give

Table 8.7 Linear correlation factors for independent variables κ, σ and three explanatory variables $Me[H_1]$, $Me[H_2]$, $Me[H_3]$—medians of the objective function values are obtained over 100 independent ESSS$_\alpha$-DM algorithm runs with the heuristics H_1, H_2, H_3. Searching space dimension $n = 30$

Objective function f_{sph}

Variables	$H_1, 2.0$	$H_2, 2.0$	$H_3, 2.0$	$H_1, 0.50$	$H_2, 0.50$	$H_3, 0.50$
κ	0.57	0.63	0.57	0.89	0.94	0.82
σ	–0.63	–0.51	–0.58	–0.28	–0.17	–0.36

Objective function f_e

Variables	$H_1, 2.0$	$H_2, 2.0$	$H_3, 2.0$	$H_1, 0.50$	$H_2, 0.50$	$H_3, 0.50$
κ	0.53	0.59	0.08	0.85	0.94	0.38
σ	0.39	0.04	0.33	–0.16	0.05	–0.18

Objective function f_R

Variables	$H_1, 2.0$	$H_2, 2.0$	$H_3, 2.0$	$H_1, 0.50$	$H_2, 0.50$	$H_3, 0.50$
κ	0.11	–0.08	–0.19	0.86	0.80	0.56
σ	–0.04	0.30	0.31	–0.15	0.26	0.31

Objective function f_A

Variables	$H_1, 2.0$	$H_2, 2.0$	$H_3, 2.0$	$H_1, 0.50$	$H_2, 0.50$	$H_3, 0.50$
κ	0.24	–0.13	–0.13	0.11	–0.26	–0.29
σ	–0.49	0.47	0.46	0.46	0.62	0.66

better results only in the case of the spherical model f_{sph}. In the case of isotropic mutations ($\kappa = 1$), this fact is confirmed by positive correlation factors for the stability index and negative for the scale parameters. It is interesting that there are opposite correlations for other testing functions. If we configure the mutation distribution to obtain frequent macro-mutations, which usually allows avoiding local optima, in this case, results become worse. Twice a weaker dependency for the stability index for f_A can be caused by the selection operator, which, under the influence of macro-mutations, looses the capacity of the directionality of the evolutionary process.

Application of tournament selection with relatively large tournament groups causes selection pressure increase, and thus only the best elements can give offspring. This is reflected in the best-obtained configurations of parameters for the ESTS$_\alpha$-DM algorithm in 30D searching spaces (Table 8.9). In most cases, the optimal dispersion κ does not strongly force the most beneficial direction. The isotropic distribution is the best exploration distribution in two cases, therefore the application of directional mutation for elitist selections is questionable. It is worth noting that, unlike in the case of algorithms with proportional selection, that optimal exploration distributions are based on the stability index $\alpha < 2$. In the case of the spherical model f_{sph}, the scale parameter is less important for distributions with $\alpha = 0.5$ than for the normal distribution $\alpha = 2.0$. The advantage of mutations based on small shifts for 2D f_e, described above, is also validated in Table 8.10. There can be observed

Table 8.8 Linear correlation factors for the independent variables α, σ and three explanatory variables $Me[H_1]$, $Me[H_2]$, $Me[H_3]$. Searching space dimension $n = 30$ and the ESSS$_\alpha$-DM algorithm

Objective function f_{sph}

Variables	$H_1, 0.01$	$H_2, 0.01$	$H_3, 0.01$	$H_1, 1.00$	$H_2, 1.00$	$H_3, 1.00$
α	–0.36	0.24	–0.40	0.34	0.32	0.32
σ	–0.47	–0.57	–0.32	–0.82	–0.87	–0.85

Objective function f_e

Variables	$H_1, 0.01$	$H_2, 0.01$	$H_3, 0.01$	$H_1, 1.00$	$H_2, 1.00$	$H_3, 1.00$
α	–0.29	0.00	–0.37	–0.45	–0.52	–0.46
σ	–0.06	–0.04	0.31	0.28	0.23	0.08

Objective function f_R

Variables	$H_1, 0.01$	$H_2, 0.01$	$H_3, 0.01$	$H_1, 1.00$	$H_2, 1.00$	$H_3, 1.00$
α	–0.39	–0.32	–0.53	–0.66	–0.67	–0.67
σ	–0.50	0.43	0.38	0.35	0.38	0.38

Objective function f_A

Variables	$H_1, 0.01$	$H_2, 0.01$	$H_3, 0.01$	$H_1, 1.00$	$H_2, 1.00$	$H_3, 1.00$
α	–0.23	–0.25	–0.25	–0.33	–0.36	–0.35
σ	–0.70	0.22	0.21	0.45	0.45	0.44

Table 8.9 Optimal parameters $\alpha^*, \kappa^*, \sigma^*$ obtained for the ESTS$_\alpha$-DM algorithm with directional mutation and different heuristics of mutation direction determination. $MNOFE$: median of the number of objective function evaluations, $MOFV$: median of the objective function value. Searching space dimension $n = 30$

ESTS$_\alpha$-DM + Heuristic no. 1

$f(\cdot)$	α^*	κ^*	σ^*	MOFV	MNOFE
f_{sph}	1.25	0.25	1	4.5129	9773
f_e	1.99	0.5	0.1	403.3957	6893.5
f_R	1.99	0.1	5	224.9375	2021.5
f_A	1.75	0.1	10	3.546	3648.5

ESTS$_\alpha$-DM + Heuristic no. 2

$f(\cdot)$	α^*	κ^*	σ^*	MOFV	MNOFE
f_{sph}	0.75	0.5	1	15.1904	9890.5
f_e	1.75	0.25	0.1	304.8702	5975
f_R	1.75	1	5	305.6067	2474.5
f_A	1.0	0.75	0.01	19.9664	11061.5

ESTS$_\alpha$-DM + Heuristic no. 3

$f(\cdot)$	α^*	κ^*	σ^*	MOFV	MNOFE
f_{sph}	1	0.5	2	25.381	9302
f_e	1.99	0.25	0.1	236.0726	6946.5
f_R	1.5	1	5	306.8533	2481.5
f_A	1.25	0.5	0.01	19.9665	10580.5

Table 8.10 Linear correlation factors for independent variables α, σ and three explanatory variables $Me[H_1]$, $Me[H_2]$, $Me[H_3]$. Searching space dimension $n = 30$ and the ESTS$_\alpha$-DM algorithm

Objective function f_{sph}						
Variables	$H_1, 0.01$	$H_2, 0.01$	$H_3, 0.01$	$H_1, 1.00$	$H_2, 1.00$	$H_3, 1.00$
α	–0.01	–0.01	–0.07	0.57	0.57	0.57
σ	–0.17	–0.55	–0.75	–0.53	–0.53	–0.53
Objective function f_e						
Variables	$H_1, 0.01$	$H_2, 0.01$	$H_3, 0.01$	$H_1, 1.00$	$H_2, 1.00$	$H_3, 1.00$
α	–0.54	–0.34	0.04	–0.02	–0.01	–0.02
σ	0.56	0.44	0.23	0.99	0.99	0.99
Objective function f_R						
Variables	$H_1, 0.01$	$H_2, 0.01$	$H_3, 0.01$	$H_1, 1.00$	$H_2, 1.00$	$H_3, 1.00$
α	0.67	0.60	0.68	0.11	0.10	0.12
σ	0.08	–0.04	0.13	–0.75	–0.76	–0.75
Objective function f_A						
Variables	$H_1, 0.01$	$H_2, 0.01$	$H_3, 0.01$	$H_1, 1.00$	$H_2, 1.00$	$H_3, 1.00$
α	–0.05	0.01	0.07	0.05	0.05	0.04
σ	0.68	0.70	0.58	0.81	0.81	0.81

Table 8.11 Linear correlation factors for the independent variables κ, σ and three explanatory variables $Me[H_1]$, $Me[H_2]$, $Me[H_3]$—medians of the objective function values are obtained over 100 independent ESTS$_\alpha$-DM algorithm runs with the heuristics H_1, H_2, H_3. Searching space dimension $n = 30$

Objective function f_{sph}						
Variables	$H_1, 2.0$	$H_2, 2.0$	$H_3, 2.0$	$H_1, 0.50$	$H_2, 0.50$	$H_3, 0.50$
κ	0.15	0.22	0.14	–0.55	–0.65	–0.58
σ	–0.77	–0.75	–0.80	0.10	0.01	–0.06
Objective function f_e						
Variables	$H_1, 2.0$	$H_2, 2.0$	$H_3, 2.0$	$H_1, 0.50$	$H_2, 0.50$	$H_3, 0.50$
κ	–0.16	–0.29	–0.37	–0.54	–0.52	–0.46
σ	0.93	0.81	0.69	0.67	0.63	0.69
Objective function f_R						
Variables	$H_1, 2.0$	$H_2, 2.0$	$H_3, 2.0$	$H_1, 0.50$	$H_2, 0.50$	$H_3, 0.50$
κ	–0.11	–0.12	–0.26	–0.40	0.01	–0.22
σ	–0.71	–0.69	–0.60	–0.39	–0.48	–0.46
Objective function f_A						
Variables	$H_1, 2.0$	$H_2, 2.0$	$H_3, 2.0$	$H_1, 0.50$	$H_2, 0.50$	$H_3, 0.50$
κ	0.16	–0.05	–0.16	0.12	–0.08	–0.14
σ	–0.38	0.76	0.71	–0.19	0.84	0.80

a strong dependence (the correlation index almost equal to 1) of the ESTS$_\alpha$-DM algorithm's effectiveness on the scale parameter σ for the isotropic distribution ($\kappa = 1$) (Table 8.11).

8.4 Summary

Simple calculation samples, presented in the first part of this chapter, emphasize important disadvantages of isotropic and symmetric distributions, as well as von Mises–Fisher $MF(\boldsymbol{\mu}, \kappa)$ directional distributions. In the first case, optimization effectiveness decreases with the space dimension increasing. In the second, deterioration in evolutionary algorithms' effectiveness is caused by an effect similar to the dead surrounding for the normal distribution. Although the density probability function of the $MF(\boldsymbol{\mu}, \kappa)$ distributions has the highest value for the most beneficial distributions of the mutation $\boldsymbol{\mu}$, this direction is chosen less and less with the space dimension increase. In order to avoid this unwanted effect, directional distributions, for which the expected direction of mutation does not depend on the space dimension, are proposed in this chapter.

The proposed directional distribution connected with a stable mutation random variable was applied to the mutation operator in the evolutionary search with soft selection algorithm. Three heuristics are used to estimate the most beneficial direction. Methodical experiments allow obtaining statistical linear correlations between particular parameters $\{\alpha, \sigma, \kappa\}$ and the adaptation heuristics applied. Results show that application of the directional distribution is thoroughly reasoned only in the case of high dimensional problems and when there is no possibility to calculate a large number of objective function evaluations. In these cases, the heuristics H_1, H_2 are most effective. They take into account not only the relative population location change, but also the fitness values of particular elements.

References

Arabas, J. (2001). *Lectures on evolutionary algorithms*. Warsaw: WNT (in Polish).

Beyer, H. G., & Schwefel, H. P. (2002). Evolution strategies–a comprehensive introduction. *Natural Computing*, *1*(1), 3–52.

Beyer, H. G., & Arnold, D. V. (2003). Qualms regarding the optimality of cumulative path length control in CSA/CMA-evolution strategies. *Evolutionary Computation*, *11*(1), 19–28.

Dhillon, I. S., & Sra, S. (2003). *Modeling data using directional distributions. Technical Report TR-03-06,* University of Texas, Austin.

Fang, K. -T., Kotz, S., & Ng, K. W. (1990). *Symmetric multivariate and related distributions*. London: Chapman and Hall.

Galar, R. (1989). Evolutionary search with soft selection. *Biological Cybernetics*, *60*, 357–364.

Hansen, N., & Ostermeyer, A. (2001). Completely derandomized self-adaptation in evolutionary strategies. *Evolutionary Computation*, *9*(2), 159–195.

Karcz-Dulęba, I. (2004). Time to convergence of evolution in the space of population states. *International Journal Applied Mathematics and Computer Science*, *14*(3), 279–287.

Mardia, K. V. (1999). Directional statistics and shape analysis. *Journal of Applied Statistics*, *26*(8), 949–957.

Mardia, K. V., & Jupp, P. (2000). *Directional statistics*. Chichester: Wiley.

Obuchowicz, A. (2003a). Population in an ecological niche: Simulation of natural exploration. *Bulletin of the Polish Academy of Sciences: Technical Sciences*, *51*(1), 59–104.

Obuchowicz, A. (2003b). *Evolutionary Algorithms in Global Optimization and Dynamic System Diagnosis*. Lubuskie Scientific Society, Zielona Góra.
Prętki P., & Obuchowicz, A. (2006). Directional distributions and their application to evolutionary algorithms. In L. Rutkowski, R. Scherer, R. Tadeusiewicz, L. A. Zadeh, & J. M. Zurada (Eds.), *Artificial Intelligence and Soft Computing*. (Vol. 4029, pp. 440–449). Lecture Notes on Artificial Intelligence. Berlin: Springer.
Rudin, W. (1976). *Principles of mathematical analysis*. New York: McGraw-Hill.
Salomon, R. (1998). The evolutionary-gradient-search procedure. In *3rd Annual Conference on Genetic Programming 1998* (pp. 22–25). University of Wisconsin, Madison.
Shao, J. (1999). *Mathematical statistics*. New York: Springer.
Törn, A., Zilinskas, A. (1989). *Global optimization*. New York: Springer.

Chapter 9
Conclusions

Technological development has been strongly inspired by solutions existing in nature since the beginning of human civilization. Creating models of the observed phenomena, we want to discover laws governing them in order to use those in designed methods and algorithms. One of the most interesting phenomena is the evolutionary process, which is an invaluable property of nature. Thanks to this process, very complex organisms, which are adapted to almost all, including extreme, conditions existing on Earth, have been created. The power of evolution lies in its rules of concurrent processing and soft selection, which, formulated in the algorithmic form, are one of the most effective tools of computational intelligence.

Evolutionary algorithms are one of the most useful methods of solving global optimization problems, both discrete and continuous. Every year, additional examples of effective evolutionary algorithm implementations in technical and technological problems are published. The main disadvantages of these algorithms are their high time and storage effort. Therefore, they are suggested to be 'last chance' algorithms, applied when classical optimization methods fail.

A class of evolutionary algorithms based on a real vector representation of an element and mainly dedicated to optimization and adaptation in multidimensional space problems is considered in this book. Among others, three important problems connected with application of evolutionary algorithms should be pointed out:

- problems with striking a balance between exploration and exploitation abilities of evolutionary algorithms,
- existence of a dead surrounding around a base point for multidimensional exploration distributions,
- weak effectiveness in problems of very high dimensionality.

This book attempts to diminish the above disadvantages by substitution of the normal distribution, most often applied in the mutation operator, by distributions belonging to the α-stable class. This seemingly small modification radically changes rules governing an evolutionary process, and strongly influences the effectiveness of optimization algorithms.

A. Obuchowicz, *Stable Mutations for Evolutionary Algorithms*,
Studies in Computational Intelligence 797,
https://doi.org/10.1007/978-3-030-01548-0_9

The lack of the analytical form of the probability density function of α-stable distributions $S_\alpha(\sigma, \beta, \mu)$ is, in the general case, the cause of poor research interest. The above limitation is weakened by pseudo-random number generators developed for the whole class of α-stable distributions (Nolan 2007). A significant disadvantage of thedistributions considered is infinite values of the variance for $\alpha < 2$ and the expectation value for $\alpha \leqslant 1$. Therefore, one can suppose that, if stable distributions with a low value of α are applied in the mutation operator of evolutionary algorithms, then the probability of macro-mutations is expected when the convergence to a local extremum gets worse. But the analysis of the ordered statistics of stable random samples shows that there exists an expectation value of some first variables of the ordered statistics if the only appropriate number λ of offspring of a base point is guaranteed (Theorem 3.11). Moreover, if λ is sufficiently large, then the expectation value of the first variable of the ordered statistics is smaller for lower stability indices α. Thus, against the previous intuitive conclusion, there is a chance for more precise local extremum localization using mutation based on a heavy-tail distribution instead of the normal one.

Whilst in the one-dimensional case the definition of stable mutation is unambiguous, there are a few of its multidimensional versions. The basic two are non-isotropic and isotropic stable mutation. The former, based on the non-isotropic symmetric α-stable distribution $NS\alpha S(\boldsymbol{\sigma})$, is obtained by adding a random variable of the symmetric α-stable distribution $S\alpha S(\sigma)$ to each component of a base point. It is a simple generalization of normal mutation. The effectiveness of mutation in stochastic global optimization algorithms has been proven by Lee and Yao (2004). However, two effects: the choice of a reference frame (the symmetry effect) and the existence of a dead surrounding, the area around the base point where it is almost impossible to locate an offspring, have strong, usually negative, influence on algorithm efficacy (Obuchowicz and Prętki 2004). The first effect makes the algorithm's exploration abilities dependent on deciding if the preferred improving direction is parallel or not to the axis of a reference frame. The second one influences the precision of optimal point localization because the radius of the dead surrounding rapidly increases with the searching space dimensionality and the stability index decreasing.

Application of isotropic stable mutation in evolutionary algorithms eliminates the symmetry effect in a natural way. But the dead surrounding one still applies. It is surprising that the range of the dead surrounding decreases with the stability index α decreasing. This is quite opposite to the non-isotropic case. One can expect that algorithms with soft selection and isotropic stable mutation with low values of α localize an extremum more precisely. On the other hand, heavy tails guarantee a higher probability of macro-mutations, so there is a chance that the exploration and exploitation abilities of an evolutionary algorithm are balanced on a satisfactory level. Additional advantages of evolutionary algorithms with isotropic mutation based on heavy-tailed distributions is their robustness to the choice of distribution parameters, especially the scale parameter σ, in comparison to Gaussian mutation. However, the application of Gaussian mutation with an optimally chosen set of control parameters results in better convergence of the evolutionary process to a local extremum.

Studies concerning an adaptation strategy for the scale parameter σ of isotropic stable mutation indicate the necessity of dependence of this strategy on the stability index α and characteristics of an objective function. Because heuristic knowledge about an adaptation landscape is usually marginal or none, this choice is unrealized in practice. Application of stable mutation based on a discrete spectral measure can be an alternative. This approach allows, additionally, modelling complicated stochastic dependencies between decision variables. So, one can fit a multidimensional stable model, used for searching space exploration, to an environment based on information included in the distribution of the best elements of the population. Preliminary studies described in Chap. 6 clearly show the advantage of this method over solutions which have no possibility to take into account possible strong correlations between decision variables. DSM weight estimation needs time-consuming local optimization in every iteration of the evolutionary process. Another problem is the choice and number of DSM spanning vectors. Solutions to these problems should be ascribed to application of heuristic approaches, which will be the goal of our future research.

Isotropic mutation based on the α-stable generator is an alternative of isotropic stable mutation. This mutation allows minimizing the dead surrounding effect on the precision of local extremum localization. The $S\alpha SU(\sigma)$ distributions are not stable, in general, but they do not give way to the $IS\alpha S(\sigma)$ distribution in applications to mutations in evolutionary algorithms. Especially mutation of the $S\alpha SU(\sigma)$ distribution with low values of the stability index in evolutionary algorithms with soft selection, and which guarantees a sufficient number of offspring of a given well-fitted parent element, is very effective in global optimization tasks. In this case, relatively good convergence to a local extremum with a satisfactory precision of its localization is obtained, and there are frequent macro-mutations which improve exploration abilities of an algorithm.

All of the above-mentioned approaches to mutation in evolutionary algorithms which use α-stable distributions, like in the case of practically all stochastic optimization methods, significantly lose their effectiveness with increasing the searching space dimension. The increase is connected with a rapid decrease in the probability of offspring location in the direction of better fitness function values. This effect especially manifests itself in the case of the commonly used symmetric mutations, including isotropic ones. The attempt to undermine this effect by application of directional mutations is presented in Chap. 8. Unfortunately, the popular directional von Misess–Fisher distributions do not meet the expectations. Therefore, directional distributions, for which the expected direction of mutations does not depend on the searching space dimension, are proposed. Results of the studies carried out show that the application of the directional distribution is justified only in the case of very high dimensional problems and when a problem specification does not allow evaluating a large number of sample points. In other cases, isotropic mutation is more profitable.

One of the most important conclusions of the famous Wolpert and Macready 'no free lunchtime' Theorem (Wolpert and Macready 1994) is the lack of the universal probabilistic model for stochastic global extremum searching. The basic conclusion of the research presented in this book is the fact that the proper choice of the probabilistic model of the mutation operator for a currently solved optimization problem is a crucial factor. Application of α-stable distributions allows obtaining more flexible evolutionary models in comparison to those with the normal distribution. Theoretical analysis and simulation experiments presented in this book were selected and constructed so that the most important features of the examined mutation techniques based on α-stable distributions can be exposed. The author hopes that the presented conclusions allow deeper understanding of evolutionary processes with stable mutations and encourage readers to apply these techniques to real engineering problems.

References

Lee, C. Y., & Yao, X. (2004). Evolutionary programming using mutation based on the Lévy probability distribution. *IEEE Transactions on Evolutionary Computation*, *8*(1), 1–13.

Nolan, J. P. (2007). *Stable distributions—models for heavy tailed data*. Boston: Birkhäuser.

Obuchowicz, A., & Prętki, P. (2004). Phenotypic evolution with mutation based on symmetric α-stable distributions. *International Journal on Applied Mathematics and Computer Science*, *14*(3), 289–316.

Wolpert, D., & Macready, W. (1994). The mathematics of search. Technical report No. SFI-TR-95-02-010. Santa Fe: Santa Fe Institute.

Appendix A
Saddle Crossing Problem

The saddle crossing problem in this book is formulated as follows. Let us consider the sum of two Gaussian peaks

$$\Phi_{sc}(\boldsymbol{x}) = \exp\left(-5\sum_{i=1}^{n} x_i^2\right) + \frac{1}{2}\exp\left(-5\left((1-x_1)^2 + \sum_{i=2}^{n} x_i^2\right)\right), \qquad \text{(A.1)}$$

where n is the dimension of the search which is, in this case, an infinite $\mathbb{R}^n$ space. The function $\Phi_{sc}(\boldsymbol{x})$ is composed of two Gaussian peaks. The lower one has its extremum in the point $(1, 0, \ldots, 0)$. The global extremum is localized in the point $(0, 0, \ldots, 0)$ (Fig. A.1).

The starting point, around which the initial population is generated, is localized in the lower local extremum of the function $\Phi_{sc}(\boldsymbol{x})$. The searching process is stopped when at least a half of a current population is located in the highest peak. In practice, when the mean value of the objective function $\langle \Phi_{sc} \rangle$ taken over all elements in the current population is higher than th value of the lower local extremum,

$$\langle \Phi_{sc} \rangle > \phi_{lim} > \Phi_{sc}(\boldsymbol{x}_0^0) \quad \text{for most} \quad k \quad \text{of} \quad k = 1, 2, \ldots, \eta, \qquad \text{(A.2)}$$

then we deal with a successful algorithm run. When the process is not successful in a given maximal number of iterations $t_{\max}$, then it is broken (Fig. A.2).

A. Obuchowicz, *Stable Mutations for Evolutionary Algorithms*,
Studies in Computational Intelligence 797,
https://doi.org/10.1007/978-3-030-01548-0

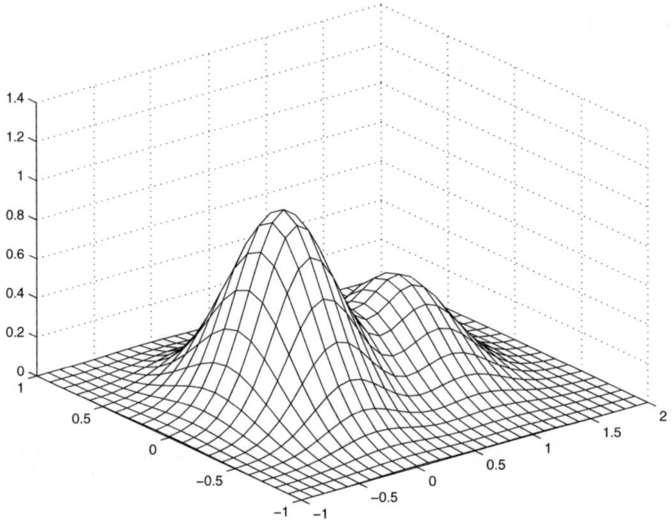

Fig. A.1 Two-dimensional version of the sum of two Gaussian peaks (A.1)

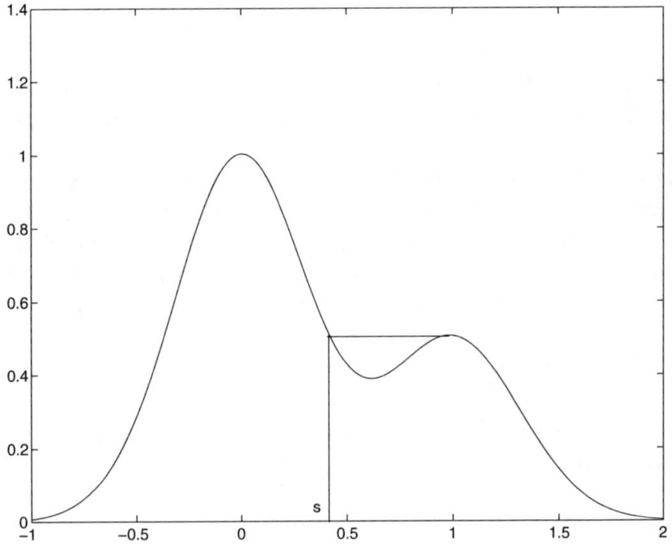

Fig. A.2 Reference point of saddle crossing

Appendix B
Benchmark Functions

All benchmark functions below are defined on an infinite $\mathbb{R}^n$ space.

- Sphere function f_{sph} (Fig. B.1),

$$f_{sph}(\boldsymbol{x}) = \|\boldsymbol{x}\|^2. \tag{B.1}$$

- Generalized Rosenbrock function f_{GR} (Fig. B.2),

$$f_{GR}(\boldsymbol{x}) = \sum_{i=1}^{n-1} \left[100\left(x_{i+1} - x_i^2\right)^2 + (x_i - 1)^2\right], \tag{B.2}$$

  where n is a searching space dimension.
- Ackley function f_A (Fig. B.3),

$$f_A(\boldsymbol{x}) = 20 + e - 20\exp\left(-\frac{\|\boldsymbol{x}\|}{5n}\right) - \exp\left(\frac{1}{n}\sum_{i=1}^{n}\cos\left(2\pi x_i\right)\right). \tag{B.3}$$

- Generalized Rastringin function f_R (Fig. B.4),

$$f_R(\boldsymbol{x}) = \sum_{i=1}^{n}\left(x_i^2 - 10\cos\left(2\pi x_i\right) + 10\right). \tag{B.4}$$

- Generalized Griewank function f_G (Fig. B.5),

$$f_G(\boldsymbol{x}) = \frac{\|\boldsymbol{x}\|^2}{4000} - \prod_{i=1}^{n}\frac{\cos(x_i)}{\sqrt{i}}. \tag{B.5}$$

- Elliptic function f_e (Fig. B.6),

A. Obuchowicz, *Stable Mutations for Evolutionary Algorithms*,
Studies in Computational Intelligence 797,
https://doi.org/10.1007/978-3-030-01548-0

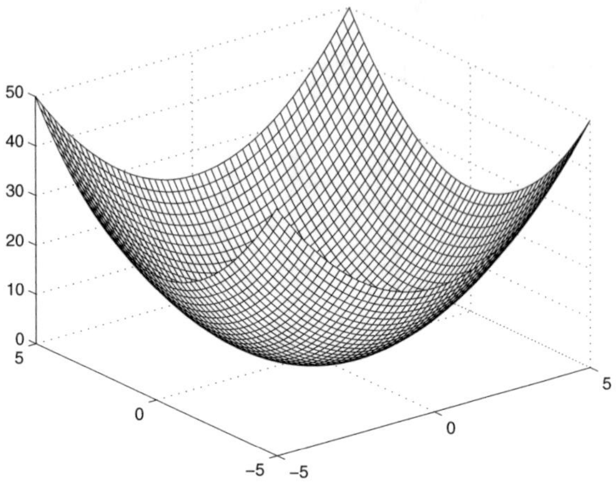

Fig. B.1 Two-dimensional version of the sphere function $f_{sph}(\boldsymbol{x})$ (B.1)

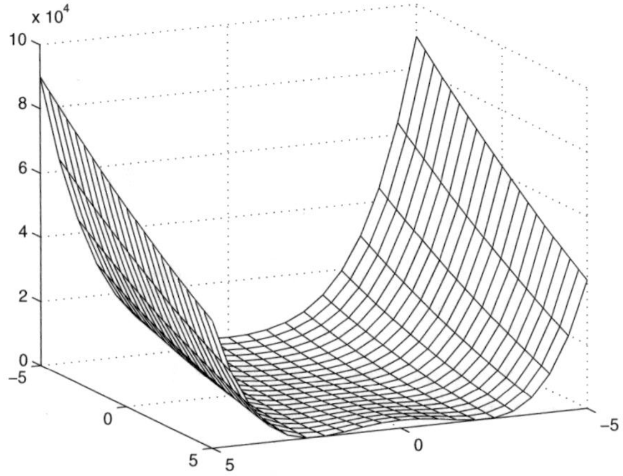

Fig. B.2 Two-dimensional version of the Rosenbrock function $f_{GR}(\boldsymbol{x})$ (B.2)

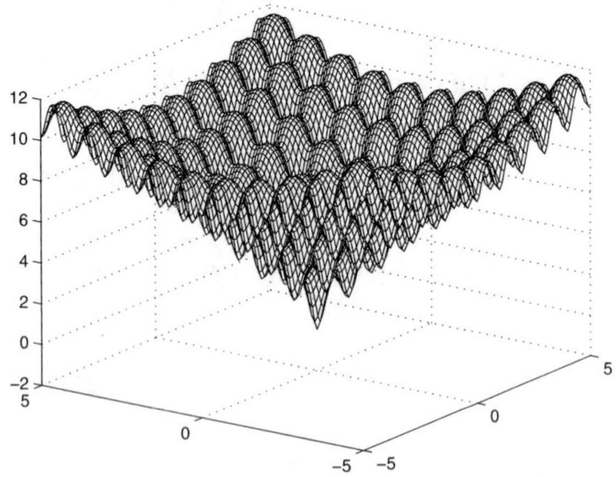

Fig. B.3 Two-dimensional version of the Ackley function $f_A(\boldsymbol{x})$ (B.3)

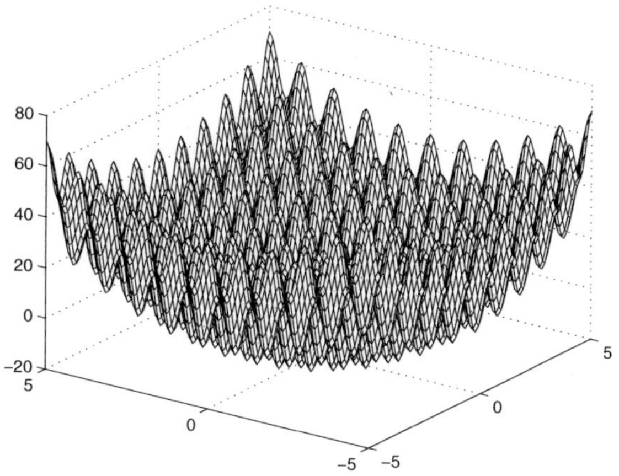

Fig. B.4 Two-dimensional version of the Rastringin function $f_R(\boldsymbol{x})$ (B.4)

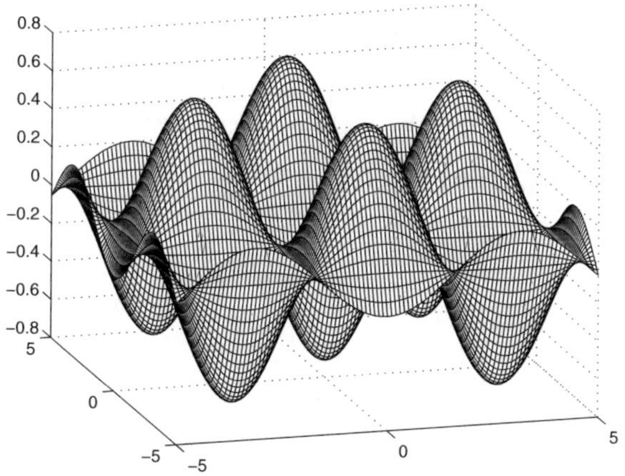

Fig. B.5 Two-dimensional version of the Griewank function $f_G(\boldsymbol{x})$ (B.5)

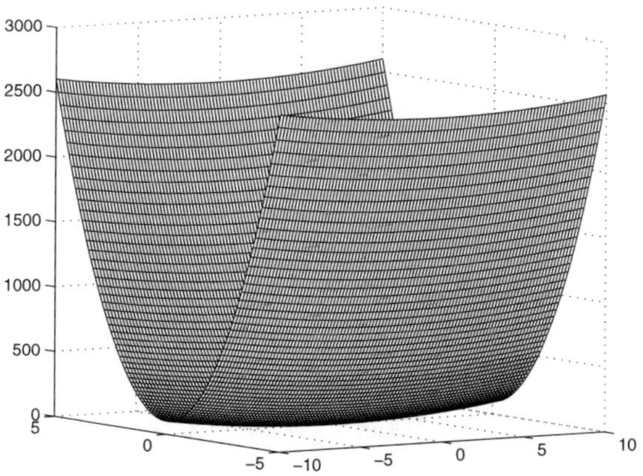

Fig. B.6 Two-dimensional elliptic function $f_e(\boldsymbol{x})$ (B.6)

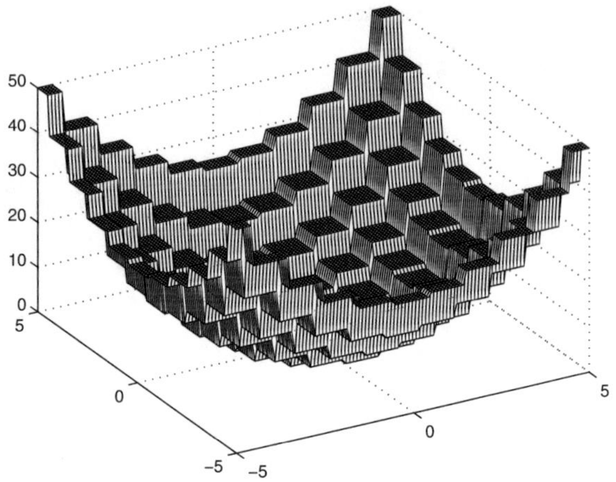

Fig. B.7 Two-dimensional step function $f_{step}(\boldsymbol{x})$ (B.7)

$$f_e(\boldsymbol{x}) = \sum_{i=1}^{n} \left(100^{\frac{i-1}{n-1}} x_i\right)^2 . \tag{B.6}$$

- Step function f_{step} (Fig. B.7),

$$f_{step}(\boldsymbol{x}) = \sum_{i=1}^{n} \left(\left\lfloor x_i + \frac{1}{2} \right\rfloor\right)^2 . \tag{B.7}$$

Printed in the United States
By Bookmasters